BUILD
BETTER JEDOX

How to build the best Jedox implementation

BY

CHRIS MENTOR

Dedication

This book is dedicated to the doers. The change makers.
The ones that turn the dial.
I have been fortunate to know many on my Jedox journey. Thank you for your energy, inspiration, and friendship.

Thank you to Dave and Vlado for review and advice.

Most of all, this book is dedicated to my family. Without you, I am nothing. Thank you for your love and support.

Contents

Introduction

Why This Book Exists

This book is intended to show you how to build well-constructed, resilient, and performant Jedox models. It is also a practical guide to get you up to speed quickly with Jedox if you are new to the technology. At the time of writing, there are no third-party books specifically on how to build Jedox applications, so this book is here to address this gap. It is not intended to be a complete technical reference guide to Jedox. There are many functions in Jedox that I have omitted, as the focus of this book is on how to start your first project. The Jedox Knowledge Base provides complete, in-depth explanations on how each component of the software stack works. It is an excellent resource, and a great companion to this book. You can access it here: https://knowledgebase.jedox.com.

Jedox is a fantastic technology, but it is often misunderstood, particularly by stakeholders outside of the Office of Finance. On the surface, it is a tool that allows organizations to streamline their planning, reporting and analytic processes.

It simplifies reporting and planning, lowers error rates in data and allows for analysis across huge datasets. But it is also so much more than just a reporting/planning application.

Jedox has been deliberately designed as a platform that contains a toolbox of incredibly powerful components. This allows users to build sophisticated models, customized to suit their needs exactly. In most Jedox projects, an initial successful implementation (typically - but not always - in Finance) often leads to Jedox being seen as the solution in more operational areas of the business, solving reporting and planning problems that no other tool can.

By the end of this book, I hope that you will have a good understanding of what tools are available in the Jedox platform, and how to use them optimally in the construction of your Jedox model. If you are familiar with Jedox already and more interested in how to use the tools to build a strong, resilient model, then I encourage you to skip directly to Chapter 7 and start reading!

WHO IS THIS BOOK FOR?

The book is written with the following people in mind:

- New users to Jedox who are about to build their first model
- Existing users looking to extend and streamline their existing implementation
- IT and Finance management personnel seeking to understand how Jedox works

Panic Notes!

If one of the following situations describe you:

- You are familiar with Jedox as a business user but have "inherited" the management and administration of the system
- You are from IT and just realized this system called "Jedox" exists in your organization, and you are now responsible for it
- You are brand new to the organization, never seen Jedox before, and now must manage it in a planning cycle

And you urgently need to work out how to address urgent system issues now, please read the **Panic Notes** chapter at the end of this book first!

> ***Note: this book is not a substitute for training or expert advice.*** *It is designed to make the you more familiar with Jedox and to show you how to create great models.*

WHAT IS JEDOX?

Jedox the company was established by Kristian Raue in Freiburg Germany in 2002, a small startup which developed an open source in-memory planning database called "PALO". There is still a small group of the original development team still at the company today, including Vladislav Malicevic who is now the CTO of Jedox.

Eventually, the technology morphed into "Jedox", a premium, paid-for planning platform.

In 2022, Jedox moved from Windows and Linux installers to a Kubernetes architecture, available as cloud-only technology. It is currently owned by US and European private equity shareholders. The company is still headquartered in Germany, with additional offices in Europe, US, Middle East, and APAC.

At its heart, Jedox has an in-memory cube-based

database engine, which makes it very efficient for not only storing huge volumes of data, but for planning and data retrieval. Jedox calculates individual user queries on the fly, allowing instant consolidation of actual and plan numbers. It uses dimensions to represent perspectives of the cube, and each dimension can contain elements that are either base, aggregated or string types. Finally, the engine has a formula-based calculation component called Rules, which allows users to build Microsoft Excel-like formulas across the multi-dimensional structure of the database.

Coupled to the database are the visualization layers of Jedox. They consist of a web design and publication area, a mobile app, and an Excel add-in. Jedox Designer on the web closely represents Microsoft Excel because the typical template designer is typically within the office of finance or a business analyst who has Excel experience. This helps to reduce the learning curve dramatically when picking up Jedox for the first time.

Finally, it's important to mention data connectivity. Jedox has easily one of the best built-in integration tools for EPM (Enterprise Performance Management) systems in the world. It employs a flexible, sophisticated Extract-Transform-Load (ETL) style of integration. This allows the user to build and schedule ETL projects connecting to a vast array of data sources, both on premise and in the cloud.

The Jedox OData Hub, adhering closely to Microsoft's widely adopted OData v4 standard, makes Jedox data available in 3rd party applications such as Salesforce, Power BI, Qlik and Tableau. If your third-party system can accept OData v4 feeds, it will be able to consume Jedox data, live.

This feature alone is a powerful argument for Jedox when working out how it fits into the broader IT strategy of an organization. It means the data is not a "black box" in Jedox; rather Jedox becomes a complimentary platform within a corporate ecosystem. Data is easily shared with other systems as required.

What is unique about Jedox? There are a lot of competitive products out there so what makes Jedox different? From a user perspective, it is ease of use and flexibility. You can model anything in Jedox. From a technology perspective, it is how it scales. Jedox can scale from a few users to thousands. Certainly, a system built for 1,000 users will be designed differently to one for 5 users, but it absolutely can be done with great design. This scaling capability means that Jedox becomes a more strategic IT investment for companies on a growth trajectory.

Although Jedox is a smaller company compared to various competitors, it has always punched above its weight in terms of its technology and its supporter base. Jedox customers and partners are long-term users who are fanatical supporters of the technology. Many features and ideas come directly from the customer and partner base. Jedox staff refer to themselves as "Jedoxians" and fervently support the company.

"WHY" OR "WHY NOT" JEDOX?

Understanding requirements is crucial to any successful implementation. Jedox is a platform that allows for complex planning scenarios, management, board and operational reporting and detailed analysis of data. It delivers this functionality through templates and reports built on the web, in Excel, or via the Jedox mobile application.

The system works best when the data you are working with is at some level of summarization and consistency. This allows you to apply standard calculations and consolidations across your dataset.

Planning is normally done at some level of consolidation - you are not normally forecasting financial data at an individual transactional level. Plans generally start at a weekly or monthly level and consolidate up from there. This level of granularity is appropriate for a planning system because it is at this level that we think about the business cycle – in weeks, months, quarters, and years. Other perspectives of the data are normally broken down to the level that makes sense for the business. For example, the sales model could contain individual products, customers or days because it is important to have that level of visibility.

Apart from heavy duty planning functionality, one of the biggest strengths of Jedox is the ability to decompose a number down to the lowest level of granularity (and then drill through to the underlying transactions for analysis), while simultaneously consolidating the data up to the highest level.

This rollup/roll down functionality has been standard in Jedox from day one, and it is still amazing how impressive this simple concept is, especially when the business is trying to grapple with multi-cost center, multi-departmental planning consolidations natively in Excel.

Coupled with this is the ability to do planning at any level in the system. Jedox always stores data at the lowest level, but you can "splash" numbers from the top level, and Jedox will automatically proportionally spread it to the lowest nodes. This gives the ability to do bottom-up and top-down planning in the same model, which provides incredible flexibility to management who are refining future forecast and plans as the business environment changes.

Not all business requirements are the same, however. If you are only looking to do reporting on detailed transactional data or produce invoices/purchase orders or other ERP related transactional reporting, then you should consider if Jedox is really the right tool for your project. Certainly, Jedox can do it - but it is not really designed for a 400-page pixel-perfect transactional style report, for example.

Jedox is an enterprise grade planning and modelling platform. It cannot be replaced currently by primarily read-only business intelligence applications, such as Power BI, Qlik or Tableau for complex planning. If you do, you will hit limitations quickly. Although these applications have some limited writeback capability, and there are indeed some excellent third-party applications that extend their capabilities, they will always have much more basic planning capability compared to what a pure-play planning tool like Jedox can do.

Glossary

Normally a glossary is at the back. I thought it would be more useful to put it up front to educate you on some terms, especially if you have not heard them before, or understand their meaning in a different context. This will help your understanding of the more complex parts in the following chapters. For an exhaustive list of all Jedox terms, please see the Jedox Knowledgebase https://knowledgebase.jedox.com.

Actions - A spreadsheet component that allows report builders to create specific actions on the spreadsheet. For example, execute an integration load or job, add elements to a dimension, or upload a file to the server.

Attributes - Are a tag of data against an element that has a one-to-one relationship with the element. Attributes can be numeric or text. For example, a customer dimension may have an attribute of "Address". Agile project management is an iterative approach to delivering a project throughout its life cycle.

Agile - A project management methodology for organizing a project. Agile life cycles are composed of several iterations or incremental steps towards the completion of a project. Iterative approaches are frequently used in software development projects to promote velocity and adaptability since the benefit of iteration is that you can adjust as you go along rather than following a linear path.

Consistency - The definition of a "common language" that defines objects in Jedox and how it is used throughout the model.

Cube - A multidimensional structure that stores data. The stored data can be stored as numbers or strings. Think of it as a database table optimized for analysis.

Database - A collection of cubes, lists and dimensions that can be organized within a Jedox instance. Objects within a single database are unique - they are not shared with other databases.

Dimension - A structure that defines a "side" of a cube. For example, a cube might contain sales data, and a dimension of that could be customer.

Dynarange - A spreadsheet component that allows the report builder to build a dynamic report that the user can interact with to change rows or columns on a report.

Element - An element is an individual component within a dimension. Elements can be string, numeric or consolidated. Elements can be organized into hierarchical structures which allow a user to see consolidated data within a cube that the dimension is attached to.

EPM - Acronym for Enterprise Performance Management, essentially a subset of the Business Intelligence industry that includes writeback, modelling and planning. Also known as CPM (Corporate Performance Management) or BPM (Business Performance Management).

ETL - Acronym for Extract, Transform and Load. Also known as Jedox Integrator. This is the data integration capability to make it easy to get data in or out of Jedox. It includes connectors that enable queries to source systems (Extract). This is followed by sorting, filtering, cleansing, and blending data (Transform), and finally sending the data to the final resting place in Jedox (Load).

FP&A - Acronym for Financial Planning and Analysis. Essentially describes the core function that Jedox assists with in an organization.

This is related to XP&A (Extended Planning and Analysis, i.e., using Jedox outside of the office of finance).

Frameset - A spreadsheet component that allows the report builder to join two or more reports into a single report. The content of each report resides in “frames” that can be dynamically changed.

Freiburg (Freiburg im Breisgau) **-** A German city in the southwest of Germany, within Germany’s Black Forest. This is where the headquarters of Jedox are based.

Jedox - The name of both the company and the technology. The founder Kristian Raue was a Star Wars fan, and it is a corruption of “Jedi”.

Jedoxian - Jedox employee and major champion of the software.

Lists - A component that allows a user to define various element intersections from multiple dimensions, the values of which can be displayed in a view on a report.

Measure - Although not mandatory in Jedox, most models contain a measures dimension. This dimension holds elements that represent a specific value that will be held there. For example, you may have a Measures dimension in a sales model that holds the elements product quantity, product margin, product cost.

Model - A model is defined here as a single or collection of cubes, dimensions and databases that are interrelated for a specific purpose. For example, a sales model might consist of 3 various cubes, plus several specific dimensions, connected by rules, templates and integrator processes. A model can be created by a user from scratch, or a model can be downloaded from the Jedox Marketplace, and installed directly on a server instance.

Multidimensionality - The concept that the Jedox database employs. That is, the ability for the user to see the same data from many different perspectives. For example, the same sales data from a sales cube could be aggregated by customer, or by region, or by quarter; depending on how the user wants to look at the data.

OLAP – Stands for "Online Analytical Processing". This describes the type of database structure that Jedox has, where data is held in cubes and dimensions instead of tables and fields.

PALO - The original name for the software when it was an open-source project. It is OLAP backwards.

Paste View - Is what is created in the spreadsheet when a user creates an arrangement of dimensions on a cube and queries the data. Cell formulas are created automatically to retrieve values from the cube for that specific view.

Rules - a formula-based calculation engine attached to cubes. Hierarchies allow users to group data from dimensions attached to cubes; rules allow a vast array of other calculations against the cube data. Rules are a key component for modelling because they are cell-based and can connect calculations between different cubes and databases. For example, rules could be used to calculate exchange rate values within a cube. You can define any area of a cube to be calculated with a rule – they are not restricted to measures.

Subsets - A collection of elements from a dimension that can be saved within the system. Subsets can be private (for an individual user) or public (for all users).

Structure - How objects are laid out or presented in Jedox.

View - a specific arrangement of dimensions and filters on a cube that can be saved. Views can be private (for an individual user) or public (for all users). Views can contain formatting and use lists and subsets.

Virtual Dimensions - Are dimensions that are not built manually, but automatically created from attribute values in a dimension. Virtual Dimensions only ever have three levels and are managed by the database.

References

- This reference section provides handy links to all things Jedox, and specific websites referred to in this book.
- The Jedox Knowledgebase - https://knowledgebase.jedox.com
- Naming Restrictions for Jedox Objects - https://knowledgebase.jedox.com/jedox/modeling/object-naming-restrictions.htm
- Jedox Academy - https://www.jedox.com/en/services/education
- Jedox Support - https://www.jedox.com/en/services/support
- Jedox Certified Partners - https://www.jedox.com/en/partner
- Jedox Downloads Page - https://www.jedox.com/en/downloads
- Jedox Design System - https://design-system.cloud.jedox.com/introduction/design-system
- The Design Interaction Foundation - https://www.interaction-design.org

Chapter One

Kicking off Your First Jedox Project

> *"Successful Jedox projects are all alike; every unsuccessful project is unsuccessful in its own way..." (Apologies to Tolstoy)*

WHAT DOES PROJECT SUCCESS LOOK LIKE?

Starting an initial project with any new software can be intimidating. There is typically a lot riding on the successful outcome of the project. However, it needn't be stressful, especially if you follow a couple of key rules. There is a lot of similarity between successful Jedox projects, regardless of scale or industry type.

SUCCESS RULE #1: INVOLVEMENT

Ensure you have Organizational Buy-In

Jedox projects that genuinely involve everyone who are stakeholders of the project are typically successful. Like most IT projects, a successful implementation is always built by a team. By stakeholders, I mean everyone who will use, consume, authorize, or manage the outcome of the project. In my experience, *this is by far the biggest indicator of success or failure of an implementation.*

A successful project typically starts right back in the buying process. End users, business analysts, the IT department, departmental and executive management must all be involved as part of the project. Ensuring everyone understands the general strategy and direction of the project and how it will help the business is crucial. The business needs to be brought along for the ride. They need to understand what is required of them. Involving people upfront *before the project has started* means that there are no ugly "surprises" when it comes to asking them for their time and resources later.

Putting a single individual in charge to design, build, manage, sign off and deliver a project without any reference points, control, support, or communication to other key personnel is generally a leading indicator of project failure.

So, what does involvement look like? Every project is slightly different when it comes to human resources, but below is a simple example of a departmental checklist for a typical, initial first phase implementation of Jedox, regardless of the size of your organization:

Example: Jedox Budgeting Project - Resource Checklist

Finance Department has personnel assigned for

- ✓ Gathering and managing requirements
- ✓ Performing data reconciliation
- ✓ Provisioning of existing templates (if any)
- ✓ Managing day to day performance of the project
- ✓ Designing of how the model will work
- ✓ Actual model building / execution

IT Department has personnel assigned for

- ✓ Setting system security, license, and infrastructure management
- ✓ Providing data integration strategy and access to source systems
- ✓ Overall project management (time/cost/resources) - note: could also be Finance

Departmental/Operations (i.e., End Users) has personnel assigned for

- ✓ Providing input into design and requirements that affect them
- ✓ Testing the templates (user acceptance testing)
- ✓ Reconciling the data that affects them

Executive team/system owner are there for

- ✓ Ensuring the cost benefit meets the expected outcome
- ✓ Understanding the overall outcome of the model
- ✓ Ensuring budget meets the scope (especially when scope changes)
- ✓ Overall ownership and final responsibility for the project

Of course, you can use your own project allocation template or software system to manage this. A team approach like the example above ensures everyone understands clearly what their job is from day one.

SUCCESS RULE #2: CLARITY

Have Clear, Realistic Deliverables

This rule is all about expectation management. This rule may seem like a generic no-brainer typical project management rule, but it is amazing how quickly a Jedox project can escalate in terms of deliverables, without consideration for the corresponding time and cost impact to the project. This escalation typically comes from early project success lifting expectations or competing / changing high priority requirements. The energy of this success can lead to unrealistic expectations and timeframes. The nature of Jedox lends itself to agile, iterative development where the project team delivers on many smaller deliverables.

Lay out what you want to achieve in the first phase with Jedox, and break that down into smaller components, and ensure all stakeholders in the Jedox are very clear on what you will be delivering on and by when. If a deliverable looks too big, or too intimidating, break it down to even smaller sub-deliverables. Below is a simple example of a high-level delivery breakup of a typical Jedox project. (Note, I use the word “sprint” arbitrarily to break up the phase into smaller components, not in the strict “Agile” project management sense of the word):

Example: Jedox Budgeting (a Finance Dept Project) – Summary of Deliverables and Outcomes. 1 Sprint = 1 Week.

Phase 1: Profit and Loss budgeting – 8 week project

- Sprint 1: Analyze existing templates, data source and calculations
- Sprint 2: Design overall model and how it will work
- Sprint 3: Build integrations into existing source systems
- Sprint 4: Create additional calculations
- Sprint 5: Build Templates, reports, and dashboards
- Sprint 6: Implement security model
- Sprint 7: User Training
- Sprint 8: User Acceptance Testing

Remember, the above is an example of how you might plan out your project at a high level. Yours could be a lot simpler or a lot more complex.

No matter how simple or sophisticated your project plan is, to aid in the success of the project it should be clear and understandable to all stakeholders.

SUCCESS RULE #3: KNOWLEDGE

Get Training and/or External Support Up-Front

If you are embarking on a Jedox project for the first time, **chances are you are not a Jedox development expert.** There is a good possibility you will require some type of assistance to get going. Do not wait until you hit challenges in the project before you reach out for help. This will put unnecessary pressure on you and the team and will potentially make any resolution much more difficult to implement. If you want assistance in your project and you have not budgeted for it, you are already in trouble: ensure that you have an approved line item on the project budget for external assistance to minimize any surprises later.

Training Before you Start

I strongly suggest that you enroll in the online (or in a classroom course) at Jedox Academy www.jedox.com/en/services/education to familiarize yourself with the various components of the Jedox platform before you start your project.

There are excellent modules there that take you through worked examples and build on prior knowledge as you progress through the course. If you are familiar with Microsoft Excel, then the Web Designer and the Excel Add-in are very easy to pick up. If you are familiar with other classical integration tools (SQL Server Integration Services, Talend, MicroStrategy, etc.) it will certainly make learning Jedox Integrator skills easier, but still relatively easy to pick up for a complete beginner.

The other components of the platform are specific to Jedox, and you want to learn about them through Academy training.

Consulting/External Support

Many implementations of Jedox start with assistance from a professional services company specializing in Jedox, or assistance from Jedox services directly. This is an excellent way to start as you can leverage off the experience and know-how of seasoned Jedox consultants. They will guide you through the project process, and you can learn a lot by working closely with them every step of the way.

Typical engagement models usually involve one or more consultants part or full time for the initial project. They then can hand off many activities to your internal team as the team becomes skilled up.

The ongoing usage of external consultants varies a lot from customer to customer. Some use consultants in the initial project to deliver quickly and on time and then revert to an internal development team for future work; other customers prefer to rely exclusively on external consultants for additional projects because of their experience and know-how and keep the internal teams for ongoing system management only. You can find a comprehensive list of certified partners on the Jedox website: https://www.jedox.com/en/partner.

If you find that you need help, do not delay – the earlier you start with external support, the better outcomes you will have with your project.

KICKING OFF YOUR FIRST PROJECT: IN SUMMARY

- **Involve your internal team**. Ensure everyone knows exactly what role they need to play and what is expected of them.
- **Ensure your deliverables are clear and understandable**. A change in requirements/scope equals deliverable change. This helps in the management of expectations.
- **Plan for Support.** Ensure some training and/or external support is budgeted in the project plan from the start and used early.

Platform Overview

In this section we will cover off the key components of the platform, in order of model building priority and how to get the most out of it, quickly.

Chapter Two

Modeler: Embody your Design

When building a new model, the first place to start in Jedox is the Modeler. The Modeler allows you to interact directly with the database level of Jedox. For those familiar with BI tools like Tableau or Power BI this is where you define your structures and calculations. However, unlike BI tools you can design the structures (the metadata) before you have loaded data.

For those familiar with relational databases, think of it as an IDE (Integrated Development Environment) like SQL Server Management Studio, PgAdmin or MYSQL Workbench for example. Here you can create all your database level objects, which is essentially your schema, without having to populate the model.

SETTING UP THE DATABASE

The first step is adding your database or databases required for your model. There are several ways to do this:

1. On the Start screen, click on "Database".

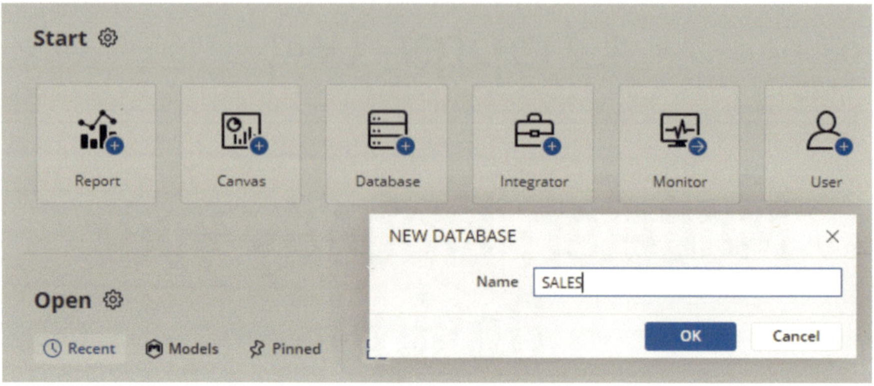

2. Click on the modeler icon in the black sidebar. Click on the blue "New database" button.

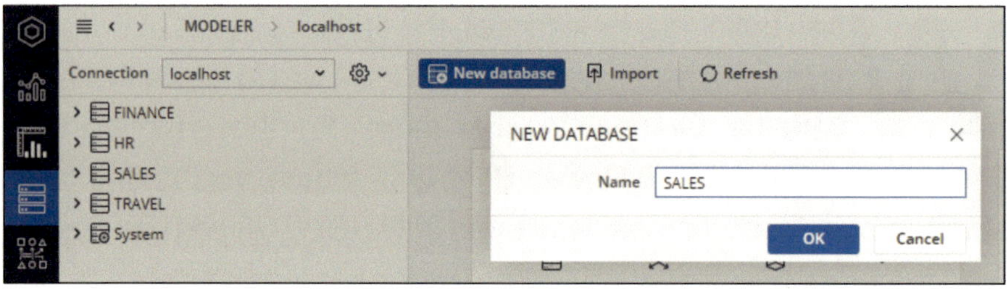

3. Click on the cog icon next to the connection name in Modeler. Then click the drop menu. Select "Create new database".

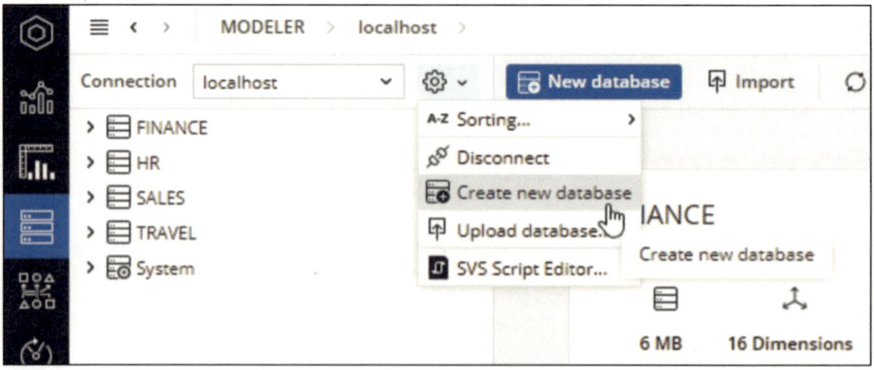

Enter your database name from your design document and click ok. This will now take you to your new database, on the Properties page. On this page you can set various security, access, and language properties. The knowledgebase has extensive information on each of these.

However, the one thing you should consider changing immediately is Visibility with user access (Right N).

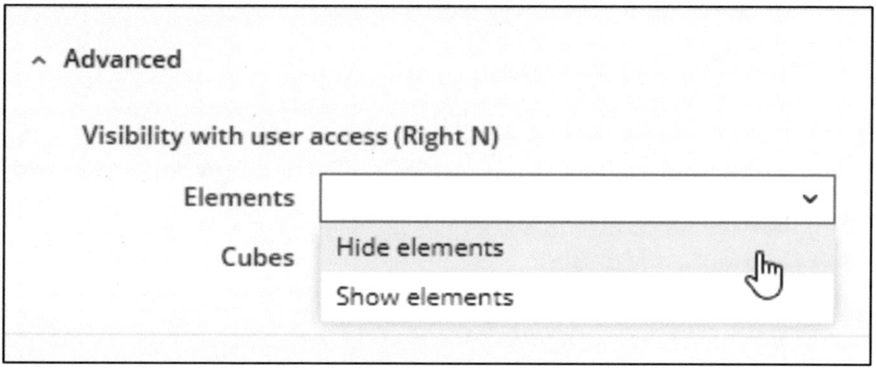

The default settings mean that a user who does not have access to a component will still see a cube or element and be able to select it, but when they try to return data against it (i.e., a paste view or a report), they will get an error instead of a value. This can be confusing. Changing this setting to "Hide elements" for elements and "Hide cubes" for cubes means that if a user has no access, these elements/cubes will be completely hidden.

For an initial build, it is best to leave the rest of the settings and set them up later during the security phase of the project.

Structure of a Database

An individual server connection contains databases which in turn contain dimensions, cubes, and lists. After creation of the database, you can see the structures underneath:

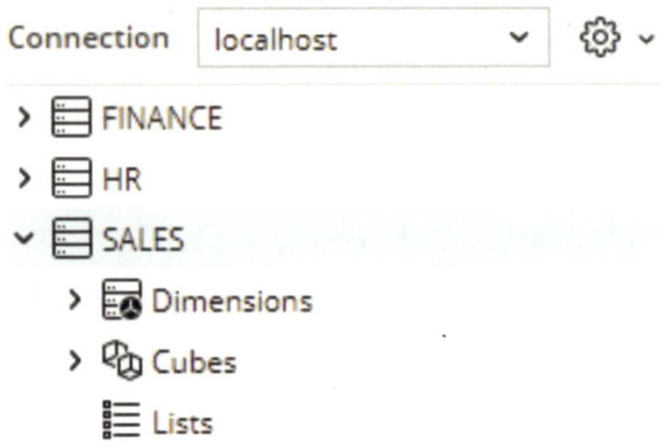

Dimensions and cubes have additional groupings for attributes, system, and rights (security) cubes. Database connections are set up in Connection Manager.

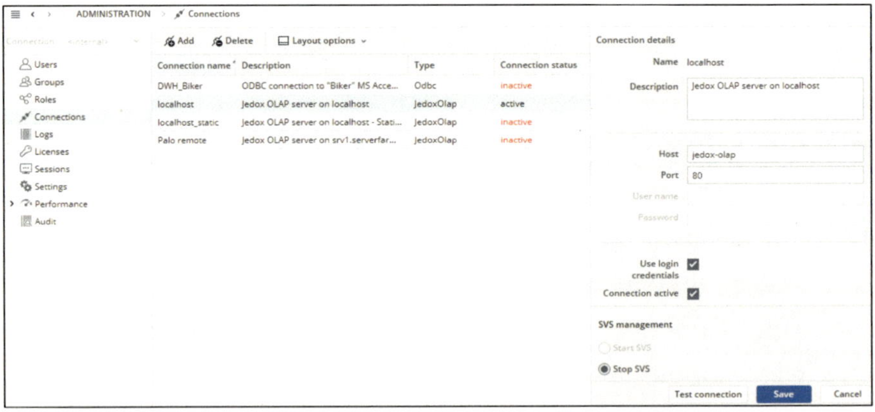

By default, Jedox provides the “localhost” connection (which is the current instance). You can of course set up additional instances in Connection Manager and view them in Modeler by changing the Connection

SETTING UP DIMENSIONS

Now that a database is established, it is possible to set up dimensions. Note, even if dimensions are to be populated by Jedox Integrator, you can still set them up manually first. This is a good idea, as it will allow you to visualize your design in Integrator before running and processes. To set up a dimension manually, right click on Dimensions, enter a name and description, and click ok.

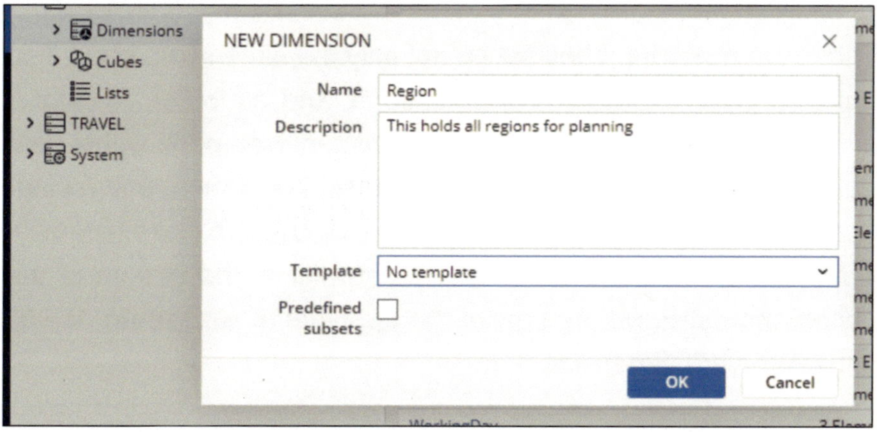

This will create an empty dimension and the screen will open to the element management screen. To add elements, simply click add element, enter the element name and press enter. The default element type is numeric, but you can change the element type to consolidated or string. To create a grouping, hover to the left of the element until the grabber appears. Then you can simply drag an element on top of another one.

Dimension Templates

Jedox has several dimension templates that you can use for standard dimensions.

Templates include:

- Version
- Month
- Day
- Year
- Legal Entity
- Region
- Account
- Account_measure
- Measure

These are powerful. They set up not only the dimension properties, but also base elements, consolidations, and attributes. To use a dimension template, right click on the dimensions node within your database. Enter the name from your design document and a clear description of what the dimension will be used for. To use a template, select an option from the template drop list at the bottom of the dialog, then click ok. As part of the template, it will create default subsets on the dimension.

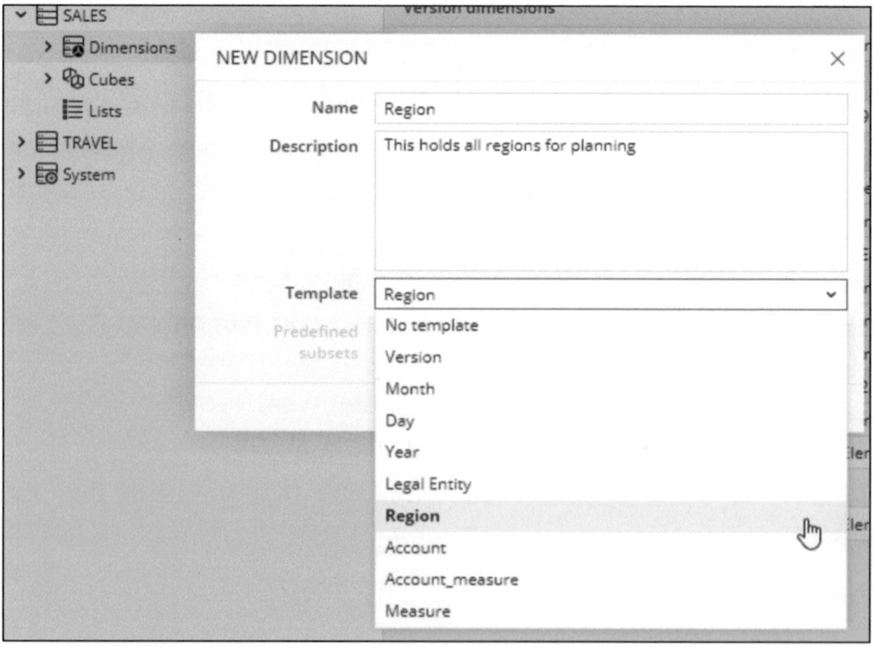

Using dimension templates where possible is a great idea, as it standardizes your model. It is also fast - you can set up several dimensions very quickly.

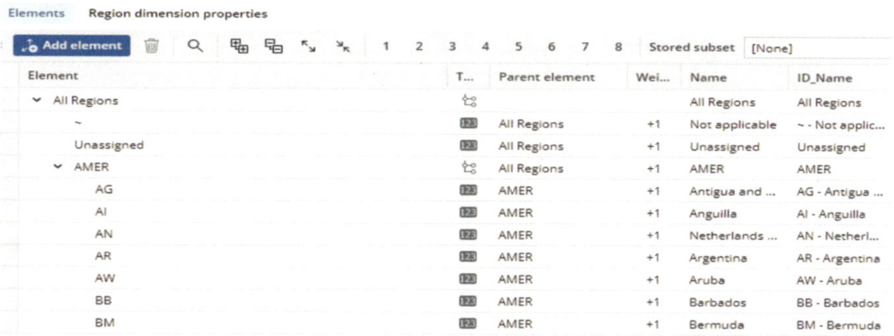

If the dimensions in your design do not have a matching template, that is perfectly fine - just give it a name and description, do not select a dimension template, and click ok. A blank dimension will be created.

Dimension Properties

Dimension properties allow you to set specific information about the dimension. There is no fixed requirement to change anything here. However, it is a good idea to at least do the following:

- **Set default elements.** Setting default elements allows the database to “fill the gaps” when information about an element in the dimension is not provided. For example, a user creates a paste view on a sales cube, but does not specify elements for the Products dimension. If you have the default element for the Total element set to All Products, All Products will be used internally by the server to return the data in the paste view.

- **Set the dimension type**. Under Advanced, you can specify the dimension type. This has 2 implications:

    1. it will classify the dimensions on the overview page by type, which can make understanding easier when you have a lot of dimensions.
    2. If the dimension is time related, setting it as a time type creates an additional tab called Time editor.

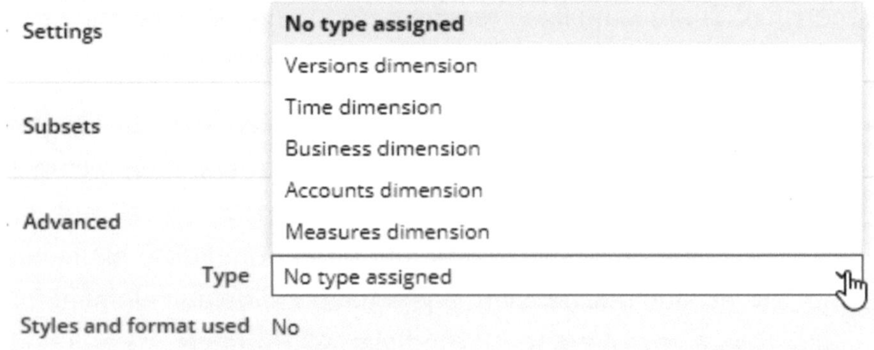

- **Add a clear description**. Think about future management of the system. It is important to document in the model exactly what specific objects do. It is helpful to have a description template that you can use.

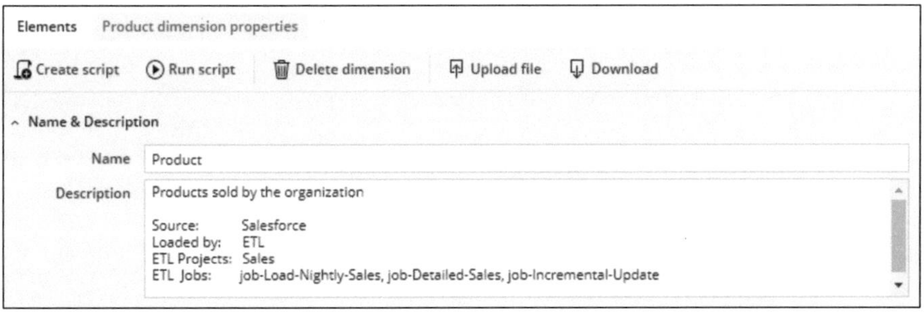

Integrator does not provide automatic references to cubes (apart from when drill through is enabled). Manually specifying any ETL jobs that update this cube provides a good audit trail of how a cube is populated/altered by integration processes.

Editor in a Time dimension

The time editor is a powerful UI to create time dimension objects. If you have used the time dimension template, default elements are created, including several helpful attributes to navigate numerically through the dimension. The dates can be adjusted by changing the date range by using the slider. Click update dimension and see the results on the Elements page. For more information on all the options available, search for Time Editor in the knowledgebase.

Creating Attributes

Attribute creation can be done on the dimension properties page or via Jedox Integrator. To create attributes via the dimension properties, expand the Attributes section and click on the "Add attributes" button. Here you can specify the following:

- **Name.** The name that the attribute will be referred to in formulas and in Integrator
- **Type.** By default, Attributes are of a string type. You can change to numeric type, if required. For example, you might require a numeric attribute to hold a percentage value.
- **Description.** A descriptive text, as it is important to explain what

the attribute is and how it will be used.

- **Domain**. If attribute entry is going to occur via the element modeler interface, you can optionally specify a Domain for manual entry. This provides data validation to ensure all entries align with the domain. For example, a year and month range Domain will allow the developer to specify a range of dates, and the user can then specify a year month combination within that range from a calendar dialog when entering in attribute values.

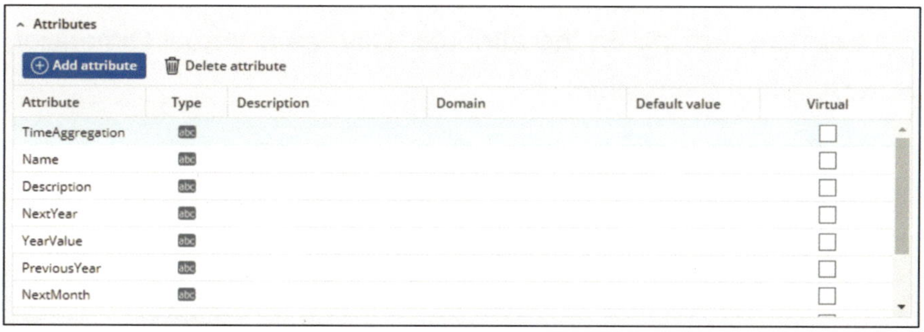

The values of the attributes can then be populated via either data entry/file upload or Jedox Integrator.

Virtual Dimensions

Once a dimension attribute has been populated, you can optionally turn that attribute into a Virtual dimension, by simply ticking the box "Virtual". This will create another dimension dynamically and will add it to any cube that uses the source dimension.

The structure of the virtual dimension is as follows:

<Attribute Name>
    Total <Attribute Name>
        Attribute Value 1
            Element 1
            Element 2

Attribute Value 2

Element 3

Element 4

No <Attribute Name> (this is where elements have no or blank attribute value)

Element 6

Invalid <Attribute Name> (this is where the attribute value is “No <Attribute Value”)

Element 7

For example, turning on the attribute “City” as a Virtual Dimension on a customer dimension:

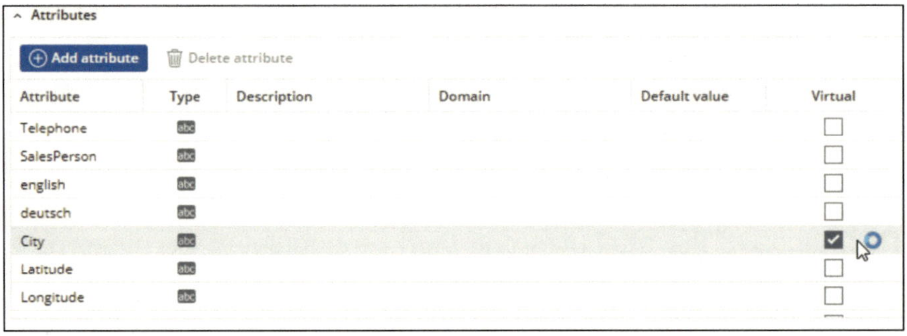

In modeler, I can see the virtual dimension has been created and is present on all cubes that use the Customer dimension (**City – Customers**):

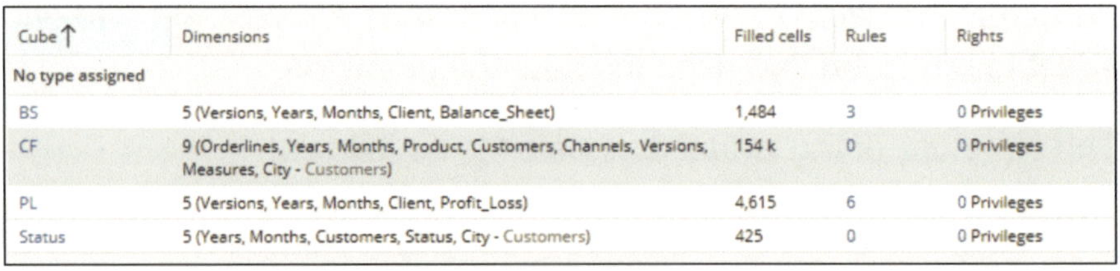

Cube ↑	Dimensions	Filled cells	Rules	Rights
No type assigned				
BS	5 (Versions, Years, Months, Client, Balance_Sheet)	1,484	3	0 Privileges
CF	9 (Orderlines, Years, Months, Product, Customers, Channels, Versions, Measures, City - Customers)	154 k	0	0 Privileges
PL	5 (Versions, Years, Months, Client, Profit_Loss)	4,615	6	0 Privileges
Status	5 (Years, Months, Customers, Status, City - Customers)	425	0	0 Privileges

In a paste view, the dimension looks like this:

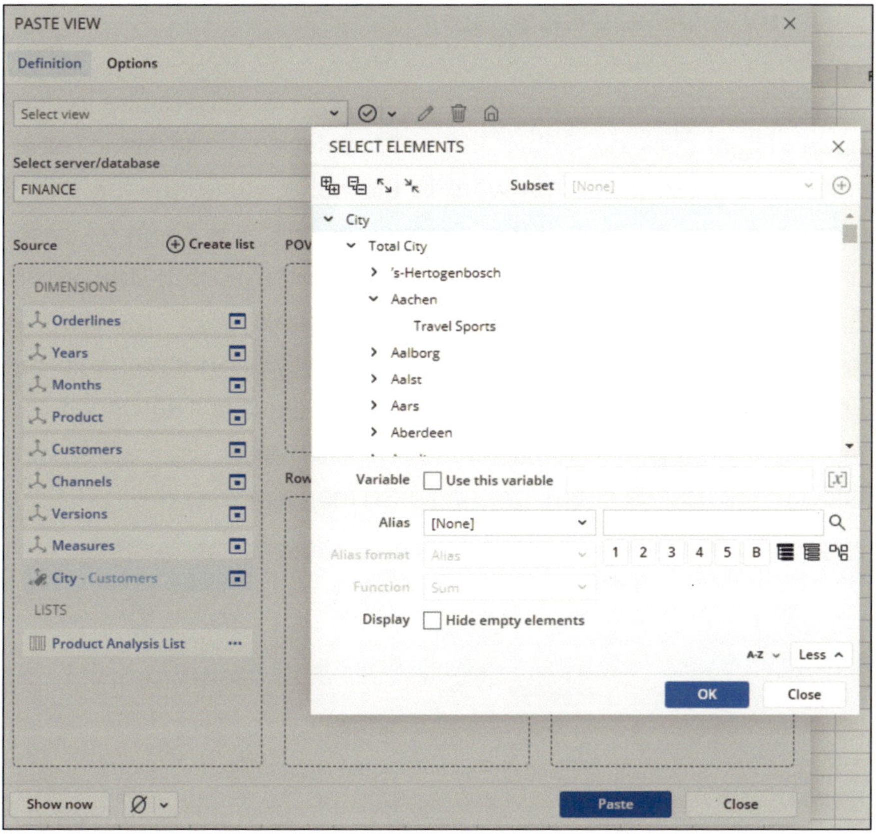

on the values entered as attribute values. If the attribute value changes, so does the hierarchy. You cannot manually change the structure. It will always be 3 levels deep. If you require a different structure, then it would be best to create the Virtual Dimension as a normal dimension, where you can have any number of levels in a hierarchy.

When you see the virtual dimension displayed on a cube in the cube properties, it will be a different color (gray/black). It will be the last dimension and the name format will be <Attribute name> - <Original dimension name>. If you hover over the virtual dimension name, it will display the system level name of the dimension.

SETTING UP CUBES

Once the dimensions have been set up, create your cubes in your model. Right click on Cubes in the Modeler tree and select New. Select the dimensions you want in the cube, remembering that it is good design to ensure the dimensions follow a common order throughout your model. If you make a mistake, you can easily rename, remove, add, or reorder dimensions after the cube is created. Do this by clicking on the “Change layout” button under the dimensions on the cube properties page. Ensure you dimension ordering is consistent across cubes and databases. This will ensure calculations across cubes are easier to manage.

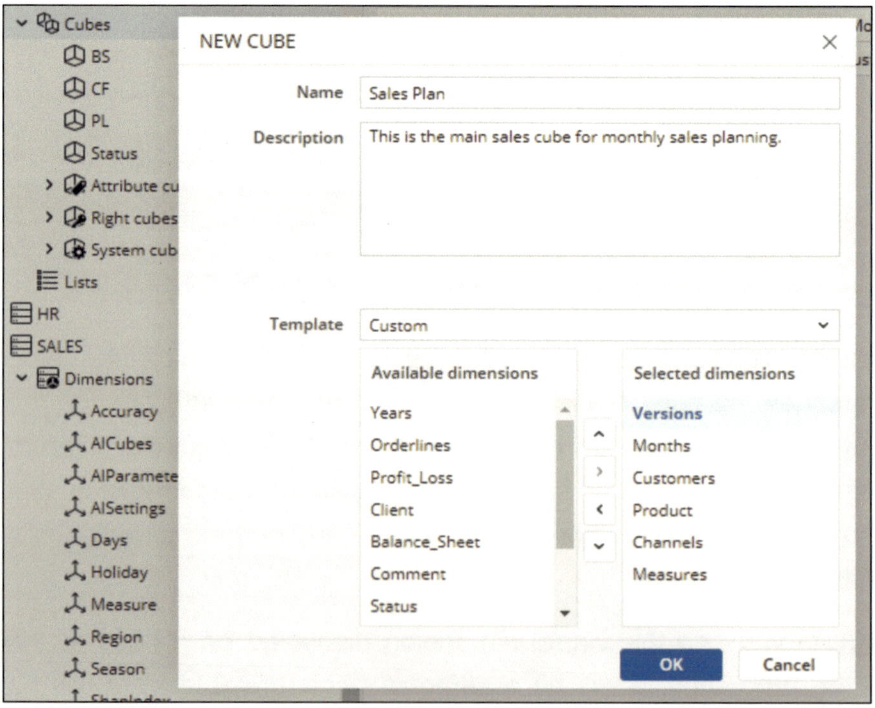

Cube Toolbar

The cube toolbar allows you to perform several actions on your cube, such as creating and running database scripts, clearing the cube of data, uploading a file to directly populate the cube, or download the cube as a file. You can also include additional business logic like version blending and KPI calculations, directly from here.

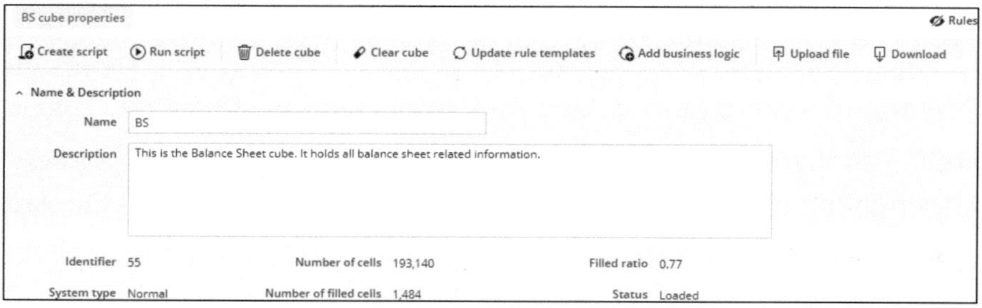

Cube Properties

An overview of some of the most used cube properties and their settings.

Description. Make the description clear and unambiguous.

Audit

This allows you to turn on cube level auditing. Once turned on, it will track every change in the cube, and record the changes (by user) internally. It will record for the specified amount of time. Integrator changes are not recorded. Logs will loop – they will overwrite and start again after the period expiry. You can view the audit logs under Audit in Administration.

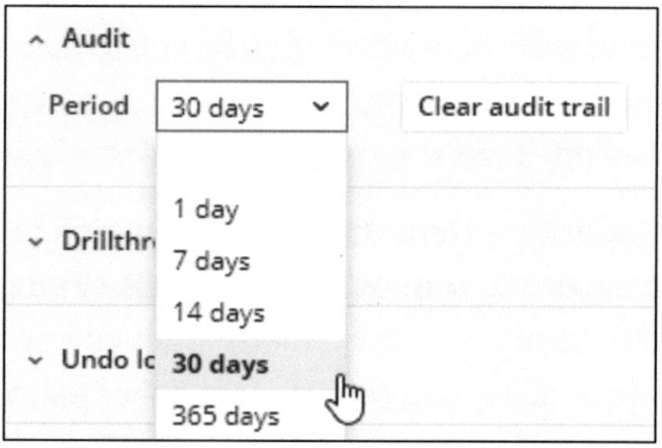

Drill through

If your cube has drill through enabled via an integration process, the information will show here. Once set up in Integrator, if you need to delete the drill through, you can reset/delete here.

Views

This shows all the public and private views that exist on the cube. Views can be created, renamed, or deleted here as well.

Advanced - Store zero values and empty strings checkbox. This is important if you need to differentiate zero cell values from blank or unpopulated cell values. By default, Jedox treats zeros and blanks/ nulls the same. By turning this on, you will be able to differentiate between them.

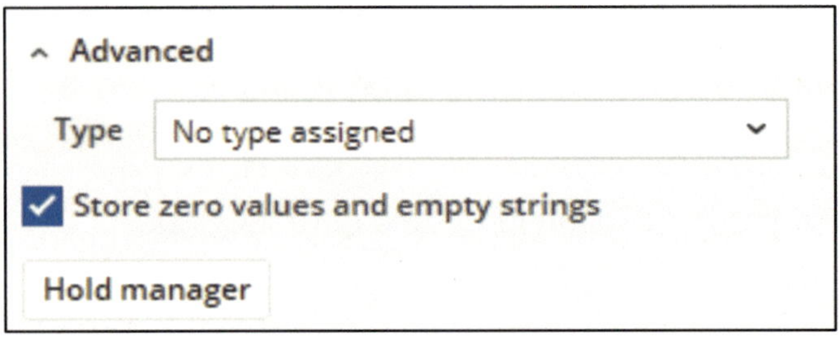

Internationalization - Allows you to hold translated string values for the cube name and description. If a user has a different language setting, it will show this information when they view Modeler.

Security – Here you can assign high level cube security. You can assign full control (admin level), Write, Read and None (blank checkbox). To assign more granular security against specific areas of the cube, you need to use either element or cell data security.

SETTING UP CALCULATIONS

Once your data structures are set up, the next thing to consider is how data will flow through your model. There are generally four types of calculations that are possible in Jedox:

- The Rules Engine
- Integrator Calculations
- Spreadsheet Formulas
- Supervision Server

Each method has strengths and weaknesses, and it is important to consider each in the overall design of your model.

The Rules Engine

Jedox has a powerful rules engine that sits in the in-memory database. It enables real time calculations on data contained in cubes and dimensions. Think of rules as Excel formulas in a multidimensional environment. Because they are like Excel formulas, anyone with Excel experience usually finds them easy to pick up. Rules can allow nearly any type of calculation within cubes, between cubes and even between databases.

ID	Position ↑	Rule	Updated
2	1	['Budget', '10 Best Bike Seller AG', 'Turnover-Merchandise'] = B:'Orders2'['Products':'All Products', 'Customers':'All Cust...	24/01/2023 09:37:09
3	2	['Budget', '10 Best Bike Seller AG', 'Cost of materials'] = B:'Orders2'['Products':'All Products', 'Customers':'All Customer...	24/01/2023 09:39:46
8	3	['Budget', '20 Beta E-Commerce Ltd', 'Turnover-Merchandise'] = B:'Orders2'['Products':'All Products', 'Customers':'All C...	24/01/2023 09:43:48
9	4	['Budget', '20 Beta E-Commerce Ltd', 'Cost of materials'] = B:'Orders2'['Products':'All Products', 'Customers':'All Custo...	24/01/2023 09:46:18
11	5	['Budget', '30 Bike Trade Sarl', 'Turnover-Merchandise'] = B:'Orders2'['Products':'All Products', 'Customers':'All Custom...	24/01/2023 09:47:57
14	6	['Budget', '30 Bike Trade Sarl', 'Cost of materials'] = B:'Orders2'['Products':'All Products', 'Customers':'All Customers', 'C...	24/01/2023 09:49:47

Rule

['Budget', '10 Best Bike Seller AG', 'Turnover-Merchandise'] = B:'Orders2'['Products':'All Products', 'Customers':'All Customers', 'Channels':'Fax/Phone/Mail', 'Versions':'Budget', 'Measures':'Sales']

Comment

Load Budget Sales into P_L cube from Orders2

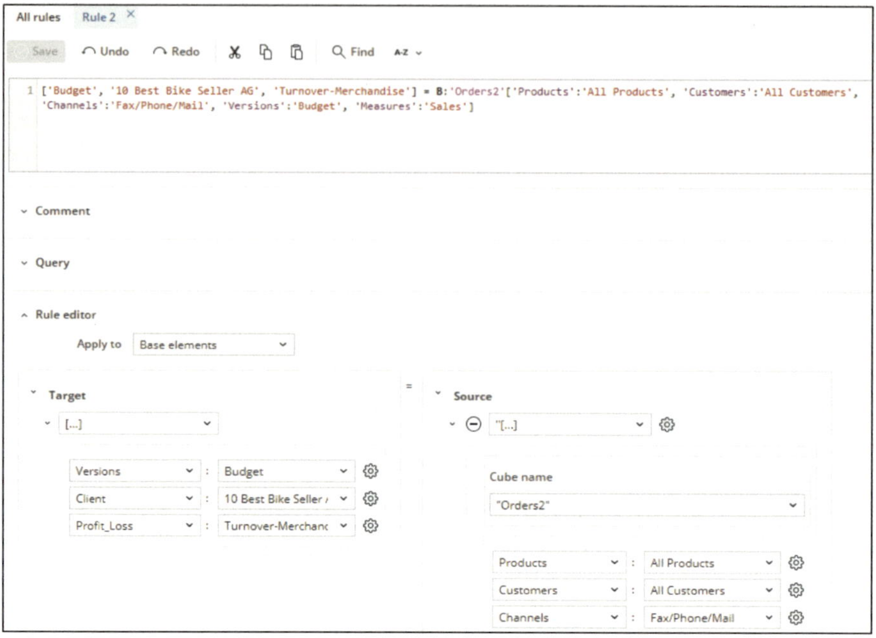

Like Excel formulas, they are volatile calculations, meaning that they are always on. For example, if a value in a cube that is used as an input for a calculation changes, the result of that calculation will change immediately.

This is very powerful and desirable in planning scenarios, as typically a user might want to change a granular value (or a driver) and see the impact immediately at the top level of the model. The implication of this volatility is that poorly designed rules can be resource intensive (depending on the size of your model) and can impact system performance.

Use rules only where required and not as a replacement for other types of calculations.

Integrator Calculations

Integrator calculations are normally static: that is, the calcs are not changing dynamically as the user data changes. However, this can have upsides as calculated cells are not continuously calculating.

Integrator calculations are normally done on the scheduled data loads against actual data. They are also done on user entered data, driven by Action buttons on templates to trigger calculations.

Integrator calculations are procedural – they start, (and potentially loop), and stop, rather than being "always on" like rules, which are declarative. For planning scenarios, Integrator is not as elegant a solution as rules; if a user forgets to trigger the calculation, values may not be calculated at that time.

Here is an example of a Switch statement which is calculating a number based upon a value of an integrator variable:

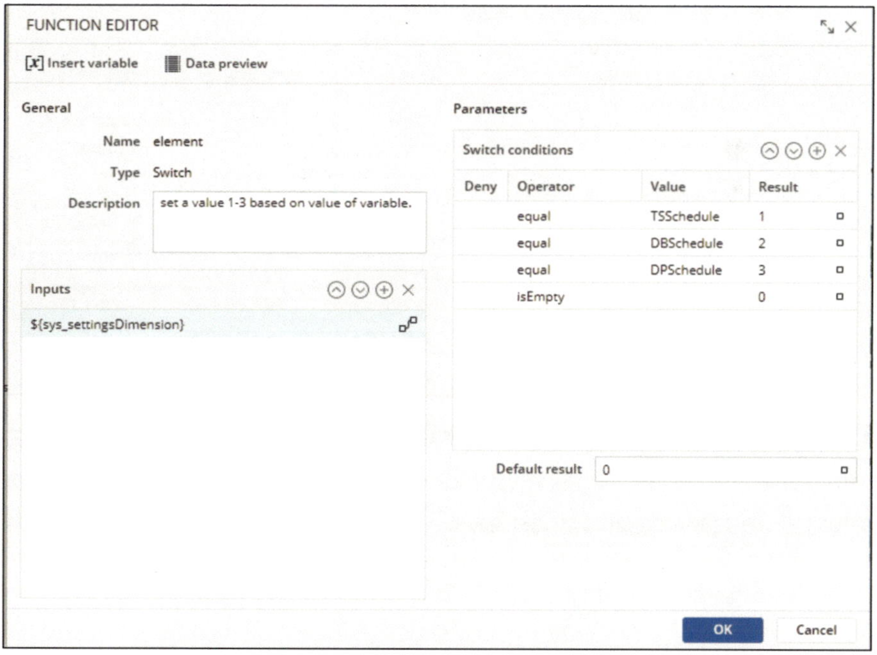

Spreadsheet Formulas

Spreadsheet formulas for calculations are lightweight and carry no calculation cost on the server as they only exist on the spreadsheet. This is an excellent solution when calculations are specific to that template and are immediate to that user.

However, because the calculations are not stored in the database, a user looking at the same dataset in Excel or in a View will not see that calculated value.

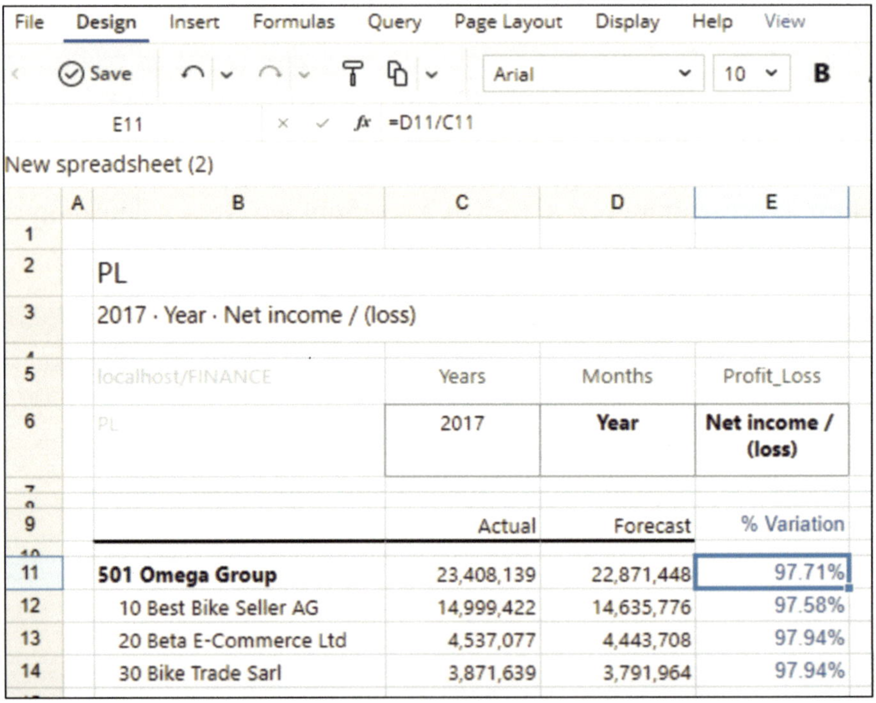

The Jedox spreadsheet on the web has an impressive number of functions available. Most major Excel functions are replicated, plus several Jedox-specific ones as well.

Supervision Server

Supervision server is a unique piece of middleware software in Jedox. It sits between the user and the database, and is event driven. Normally it is used for security authentication and authorization, but it can also be leveraged to listen to areas of a cube and perform calculations as values change based on user input.

Using Supervision Server as a calculation engine sounds great in theory as it calculates once as required (like Integrator) with no "always on" volatility (like rules). But there is a downside. Supervision Server is run via PHP scripts. These need to be written and maintained and can easily grow to thousands of lines long. When you use it over an area of the cube, a lot of the inbuilt planning and writeback functionality needs to be re-replicated in Supervision Server. PHP execution of these scripts is also not as fast as the in-memory database engine, so delays to processing may take place.

There can be specific use cases where Supervision Server is very useful in a data model, but the standard use case is for user authentication.

sep.inc.php × examples.php × sep.inc.units_positive.php ×

Save Save all

```
  public function InitCubeWorker($database, $cube) { // void
     sep_log("<< USING SAMPLE SCRIPT PLEASE ADJUST TO MATCH YOUR CONFIGURATION >>");
     sep_log("<< Get triggering area >>");
     if ($database == "Demo")
     {
        if ($cube == "Sales")
        {
             //Set the area of the database:Demo and the cube:Sales that has to be watched
             //The variable DIMENSION_TOTAL includes all elements of the dimension.
             //The dimensions follow the same order as in the modeller of the Excel addin.
             $Area_to_watch = Array(DIMENSION_TOTAL,DIMENSION_TOTAL,DIMENSION_TOTAL,DIMENSION_TOTAL,DIMENSION_TOTAL,Array('Units'));
             //The next line tells the SVS to execute a function named neagatveUnits once a cellvalue in the
             //watched area has been changed.
             $this->WatchCubeArea($Area_to_watch,'negativeUnits');
        }
     }
  }
  public function OnDrillThroughExt( $database, $cube, $mode, $arg, $sid ) { // string
     sep_log("<< USING SAMPLE SCRIPT PLEASE ADJUST TO MATCH YOUR CONFIGURATION >>");
     sep_log("<< OnDrillThrough >>");
     return "not implemented;\r\n";
  }

  public function InitDimensionWorker()
  {
     sep_log("<< USING SAMPLE SCRIPT PLEASE ADJUST TO MATCH YOUR CONFIGURATION >>");
     sep_log("<< Get triggering dimensions >>");
  }
}
```

Now that we have established databases, cubes, dimensions, and calculations, it's time to look at how to populate the data in your model.

Chapter Three

Loading Data: Getting it into your Model

There are three main methods to populate data into your model:

- Manual data entry, using PALO.DATA formulas
- Upload a file via Modeler
- Load data via Jedox Integrator

MANUAL DATA ENTRY

This is how end users will typically enter data into the model. You can also use manual entry to populate attributes, elements, and small cubes that might hold certain drivers that we might refer to later (e.g., interest or currency rates).

To add elements to a dimension, you can manually copy and paste values from Excel to Modeler. You can then manually drag and drop elements on top of each other to create a hierarchy if required.

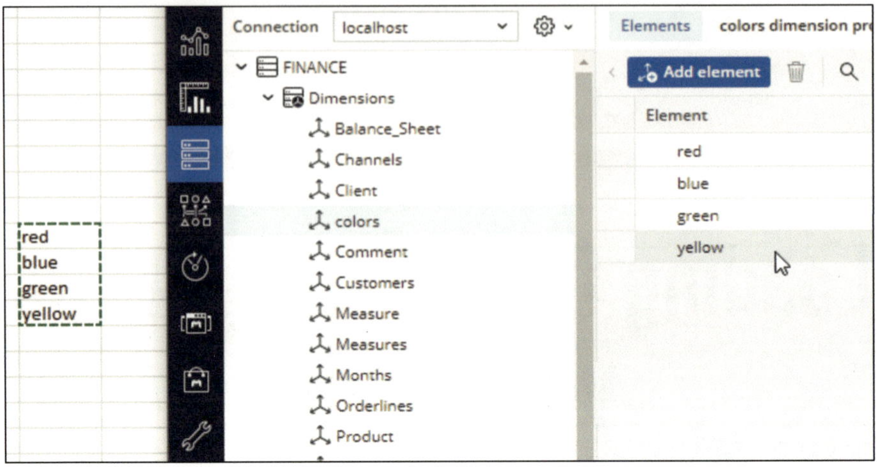

To send data values in, create a paste view on the cube. Jedox allows you to type directly over a cell to enter data at that intersection (e.g., PALO.DATA formula). Make sure you are at the base level in the cube (or use # at the front of the number to splash across consolidated values):

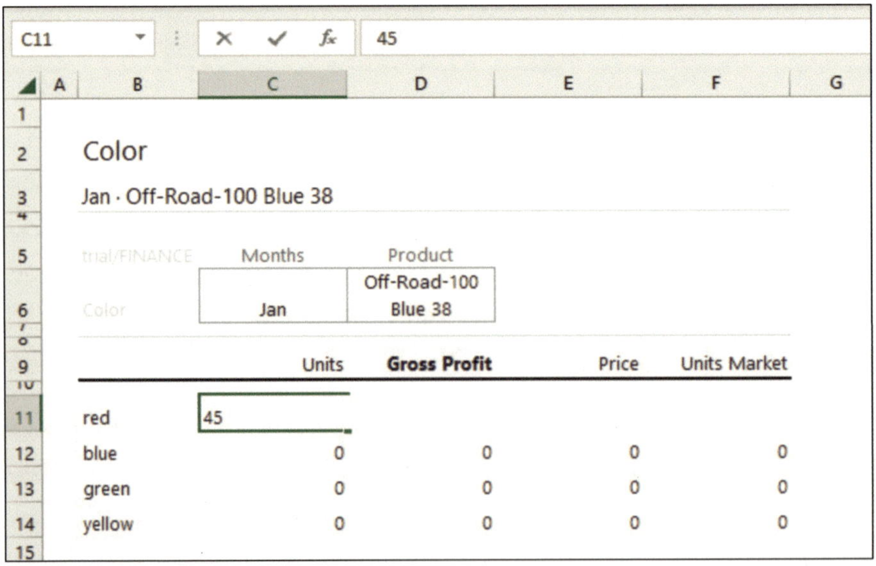

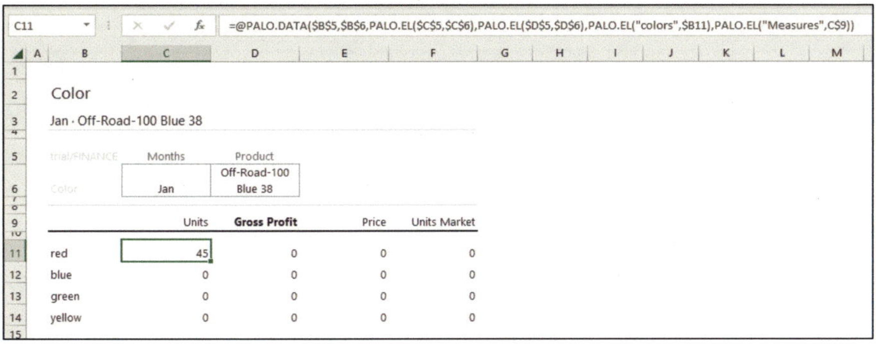

Jedox also allows you to point a cell formula to another value in the spreadsheet and send that value into the cube on sheet recalculation (e.g., PALO.SETDATAC formula).

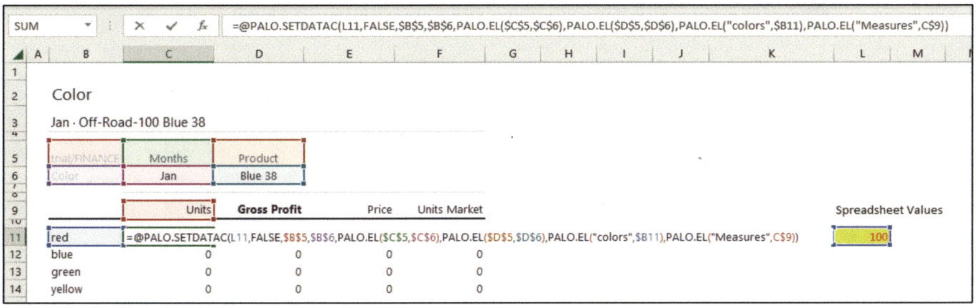

Remember, if you use a view, or use the Paste Data function, the formulas are written for you. The Knowledgebase contains examples and more information on how to use these formulas. The PALO.SETDATAC approach is an excellent way to populate a cube with data from existing spreadsheets in a manual but fast approach. This approach allows you to load spreadsheet-related data quickly and to review the data in your cubes.

UPLOAD FILE (DIMENSIONS AND CUBES)

You can right click on a dimension or cube and select the Upload file option. This will start a wizard and allow you to point to a text file and load data directly into a cube or dimension. The wizard asks you to select a file (locally or in Designer) for dimensions (Dimension Upload Wizard) or cubes (Cube Upload Wizard). Both wizards allow you to construct an upload process and optionally save the result as an Integration project. This is powerful as it allows you to repeat the process if required by just executing the Integrator job. The downside is that there are limited data transformation options in the wizard - it is really designed to get data into your model quickly and assumes that the data is already in a particular shape.

LOAD VIA INTEGRATOR

Loading cubes and dimensions via Integrator is the "industrial strength" approach as it gives you complete control over the extraction, transformation, and load processes. Let's review each of the key components of Integrator.

Variables

Variables are the key to making your integrator projects flexible. They can be used nearly everywhere in an Integrator project, including groovy script functions and jobs. By specifying variables, you can easily reuse components for different functions in your project. For example, you might have common levels and attributes for several different dimensions. Instead of creating separate connections, extracts, transforms, loads and jobs for each, you could create one chain of components with variables to create the correct dimension.

Jedox variables can be hard-coded, read from system keys or calculated via a groovy script.

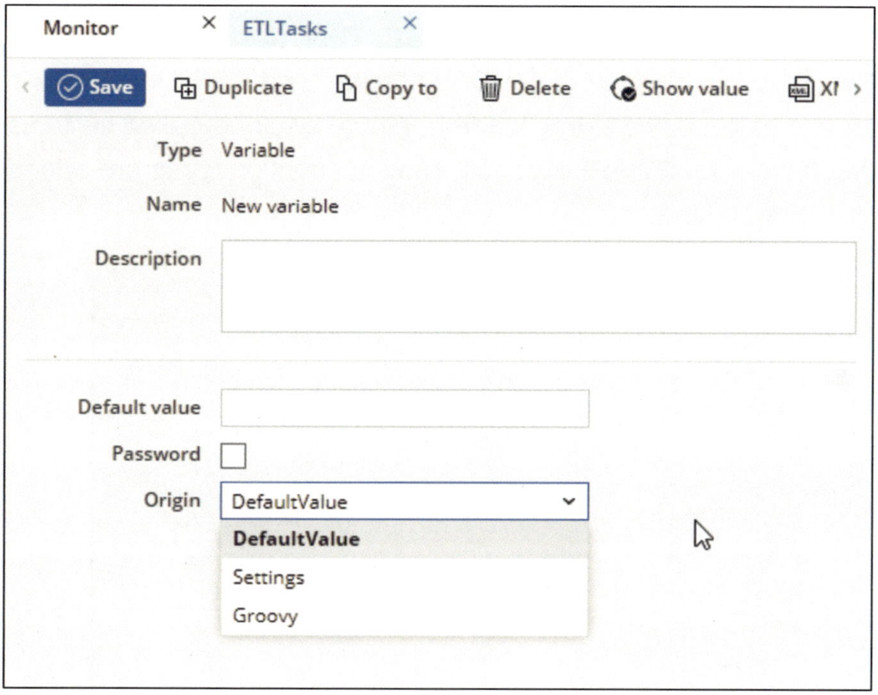

Variables are also used in loop jobs or transforms when you need to loop through elements, for example. The downsides to using too many variables is that in the absence of clear documentation, your model can become very complicated, and a new user may struggle to understand exactly what is going on in the project. Think carefully about when variables are appropriate. Use variables when it is logical and efficient to do so, while maintaining the readability of the project structure. Always document what the variables do and how they are used in the model.

Connections

Connections allow you to point to nearly any data source you can imagine and load that data into Jedox. Out of the box, Jedox can connect to over 32 specific data sources, including generic sources such as REST, XML, Generic JDBC, CSV (text) files and Excel files. The combination of the OAuth Token and REST connections for example, allows extraction from nearly all cloud-based systems that have a restful API exposed. New connection types are added from time to time.

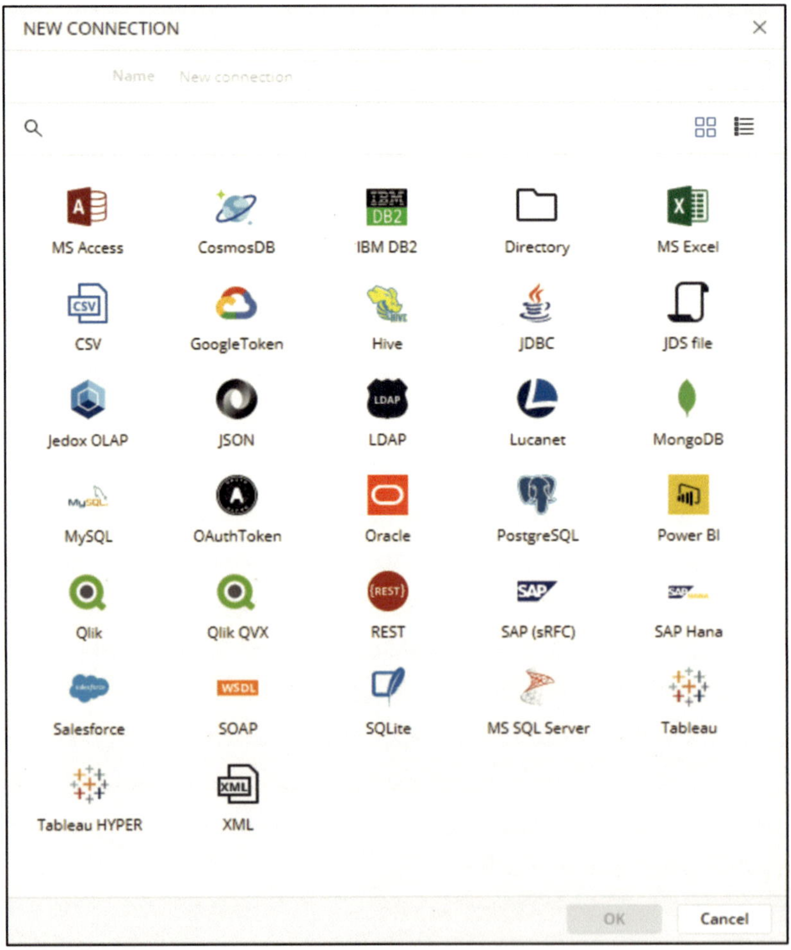

Extracts

Once a connection is established, Extracts allow you to pull the data from its source and preview the results. Depending on the data source, the extracts required might be different. For example, SAP, Salesforce, Qlik and LDAP have specific extracts defined, whereas file, Excel, JSON and XML are more generic in nature.

The Jedox extract is flexible. It allows you to have generic extracts with variables but also to be very specific, which saves on transformation steps later.

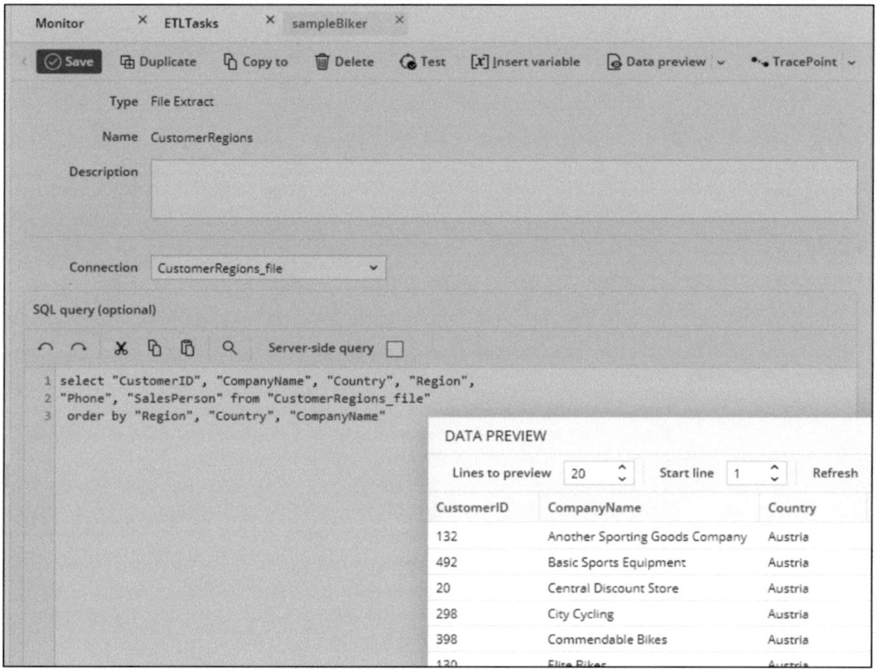

Try to use extracts to minimize the number of transformations you need. For example, in a relation extract, write a query that gives you the closest shape possible of the data you need to load. It will result in less steps in transformations, and an easier integrator project to understand.

Transforms

Transforms allow you to change the shape of the data, enhance it with lookups or joins. They are the main way to organize data for the eventual loading into cubes, dimensions, or other systems.

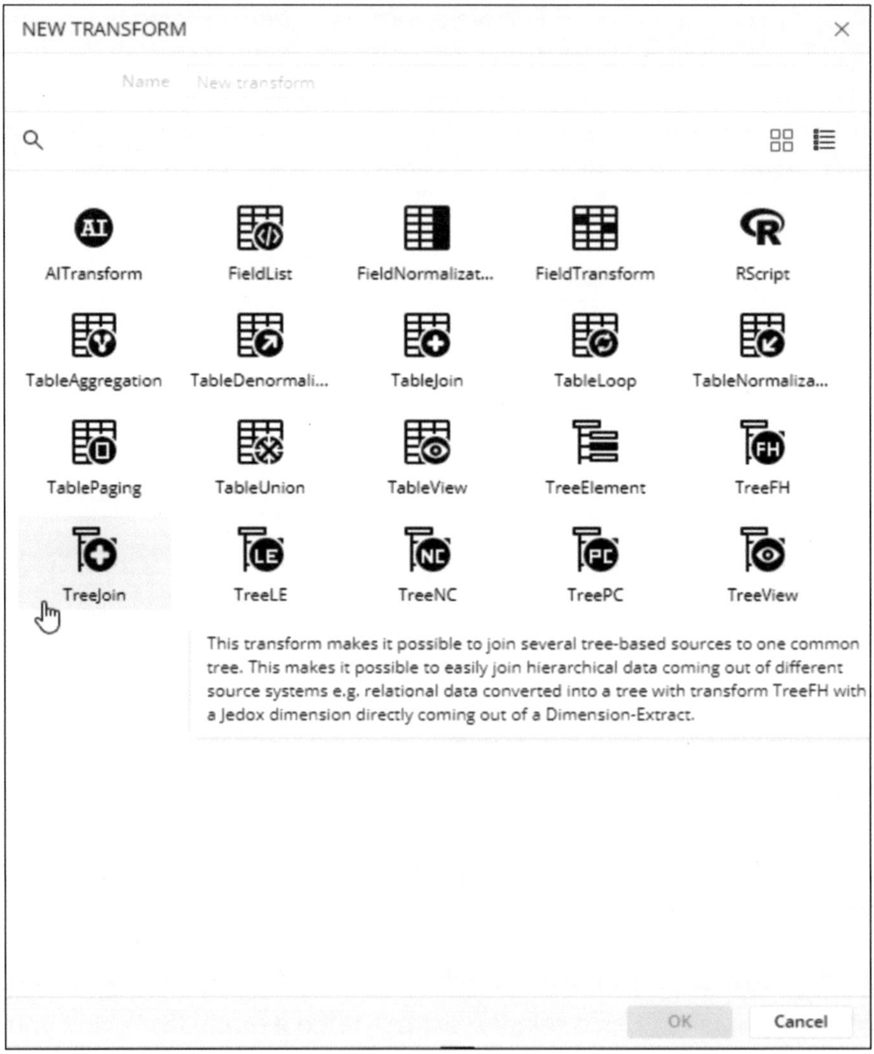

When using transforms think about how to structure them. Consider the order in which they execute and how many you need to use. Be as efficient as possible with their use, as every transform step will add additional processing time (and complexity) to your project.

Loads

Loads are the final step in a chain of components in an integrator. The load step tells the integrator where you want to send the data that you have extracted and transformed. It is a mandatory step if you want to send your data somewhere.

The flexibility of Integrator means that you can not only load cubes and dimensions, but to many other destinations as well, such as Power BI, Qlik, Salesforce and Tableau. You can execute a load directly, or as part of a job.

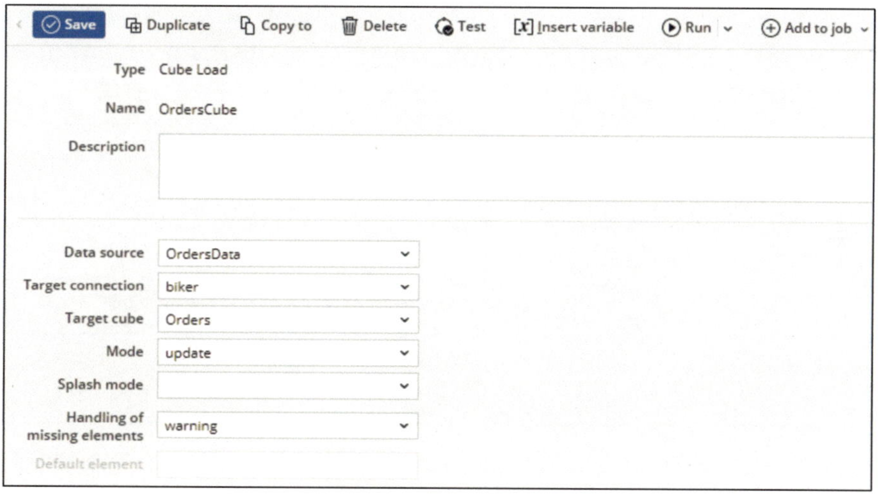

Jobs

Jobs allows groupings of loads (or other jobs) to be executed. Here you can specify the method of execution, including calling an external job (from another project), a loop, loop parallel, parallel, or switch job. In previous versions of Jedox, users would typically use a groovy job to execute anything other than a straight list of loads. For example, you could write a groovy script to execute one load when the condition is true, another when false. There are specific components to do this (in this case, the Switch Job). Always use job types where possible and only use groovy scripts where necessary. This keeps the project cleaner and easier to read.

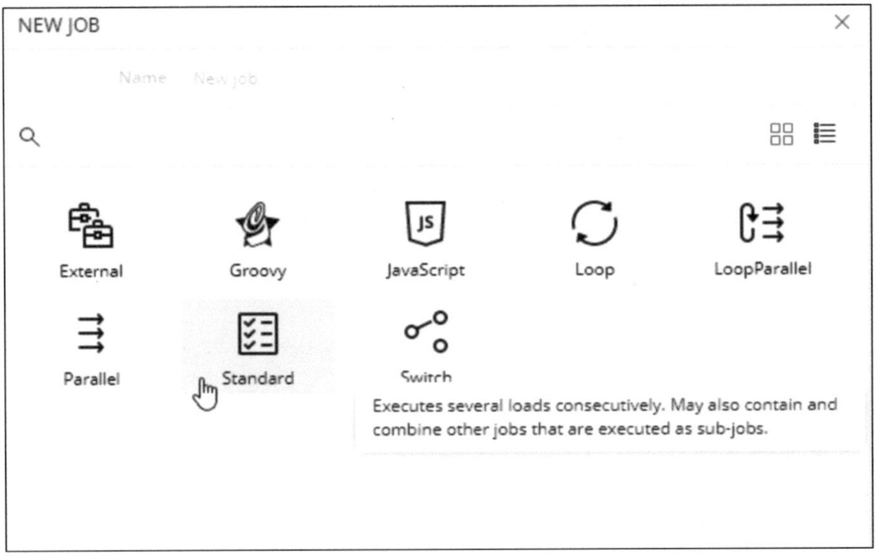

Using Flow Graph to navigate

The Flow Graph (usually on the right side of your Jedox project) shows the steps in a particular chain of components. It is a great way to visualize how a series of steps relate to each other. Hovering over the component name will show any variables that are used in the steps. Clicking on a component will take you directly to that step.

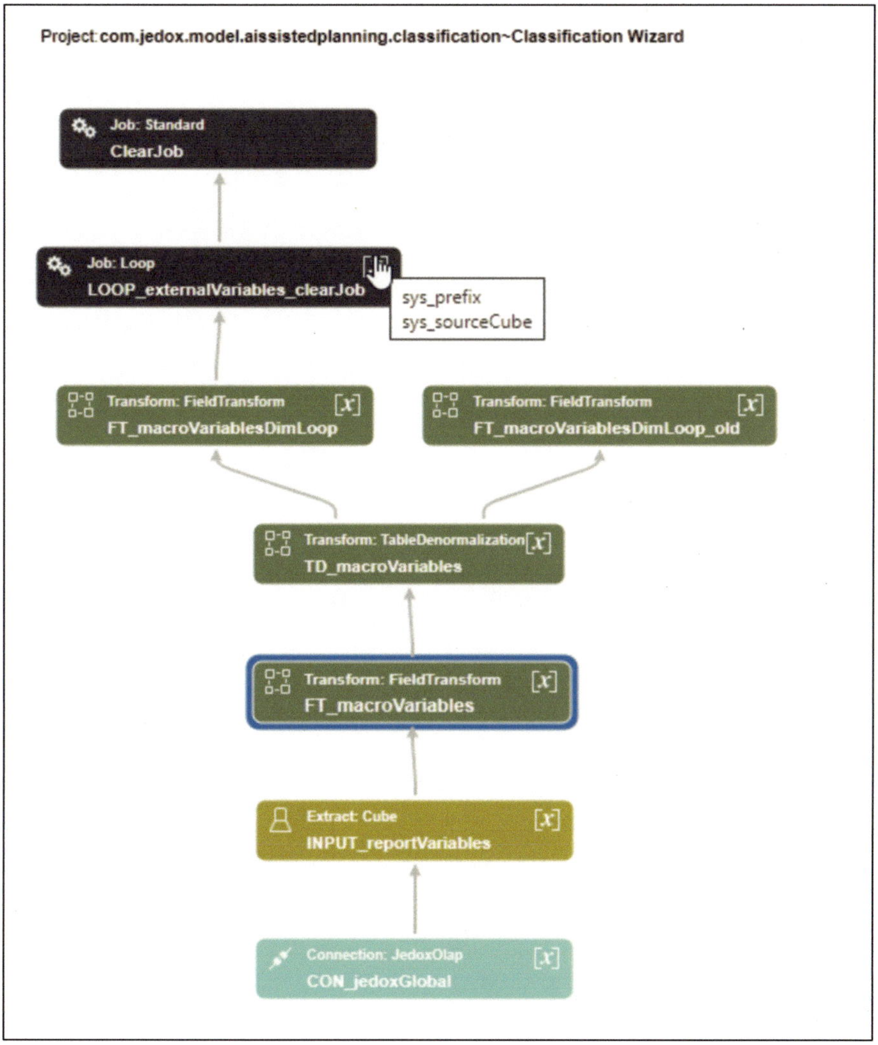

Chapter Four

Visualizing your Data

After loading, the next step is to visualize your data. There are two steps in this process - building templates in Designer and publishing the template in Reports.

The first step is to understand how your users will be consuming the data. For example, will the users be using Jedox web for all their data requirements (reporting, data entry templates, analysis) or will they be using the Jedox Excel add-in (as it is a more familiar platform), or will they be using a combination of both? There are good arguments for using the web environment or the Excel add in.

When planning templates or reports need to be distributed to many people for planning contribution, the web is normally the best approach. The web lends itself to a "planning application", where there might be a suite of reports, templates, and dashboards you want your users to consume in the easiest way possible. Additionally, you can simplify the templates, making them "tamper proof".

Where users have requirements for very large templates (hundreds of columns, by hundreds or thousands of rows), or where users have detailed analysis requirements, Excel with the Jedox add-in may be the best delivery system for your templates. It is important to get this right, as the delivery platform can have a big impact on the perception (and therefore success) of your model that you are rolling out. Typically, it is not an "either/or" decision: many Jedox projects employ both the web (Reports/Canvas and Views) and the Excel add-in, with the Excel add-in reserved for "heavy lifting" analysis duties with large spreadsheets.

Decision time: When is it best to use Jedox web for report delivery?

☑ The planning model is application-like, with multiple reports, templates, and menus

☑ The reports / templates generally fit into single screen width

☑ A large user group, where planning and analysis is not their day job

☑ The model takes advantage of web only components (e.g., Canvas, Dynaranges)

...And when is it best to consider Excel add in for reports/ template delivery?

☑ The users are primarily analysts, where they will want to see the data model in detail

☑ The reports/templates are large, with many rows or columns that require scrolling

☑ The users need to connect to existing spreadsheets to send data into Jedox

☑ The users do not want to move from the familiarity of Excel

VIEWS - POWERFUL AND QUICK

Jedox views are queries on individual cubes which allow the arrangement of dimensions on row, columns, and titles. These can be saved as either public or private (by user) and are available in Excel and web. Originally developed as an analysis tool, views have become much more report-like with the addition of themes and lists. This means that you can use views as reports or data entry templates.

The biggest advantage of views is that they are very fast to set up and easy for users to use. The biggest downside when using them as reports compared to the cell-by-cell spreadsheet is that there are limitations on form elements, formatting and layout that can be achieved within a view. You might not always be able to configure the exact format you require. However, depending on what you are trying to achieve, views may be an excellent visualization vehicle for your model.

Reporting View example:

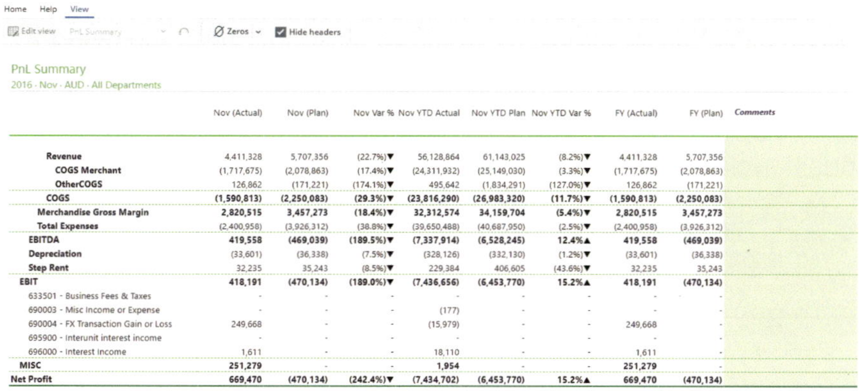

PnL Summary
2016 · Nov · AUD · All Departments

	Nov (Actual)	Nov (Plan)	Nov Var %	Nov YTD Actual	Nov YTD Plan	Nov YTD Var %	FY (Actual)	FY (Plan)	*Comments*
Revenue	4,411,328	5,707,356	(22.7%)▼	56,128,864	61,143,025	(8.2%)▼	4,411,328	5,707,356	
COGS Merchant	(1,717,675)	(2,078,863)	(17.4%)▼	(24,311,932)	(25,149,030)	(3.3%)▼	(1,717,675)	(2,078,863)	
OtherCOGS	126,862	(171,221)	(174.1%)▼	495,642	(1,834,291)	(127.0%)▼	126,862	(171,221)	
COGS	**(1,590,813)**	**(2,250,083)**	**(29.3%)▼**	**(23,816,290)**	**(26,983,320)**	**(11.7%)▼**	**(1,590,813)**	**(2,250,083)**	
Merchandise Gross Margin	**2,820,515**	**3,457,273**	**(18.4%)▼**	**32,312,574**	**34,159,704**	**(5.4%)▼**	**2,820,515**	**3,457,273**	
Total Expenses	(2,400,958)	(3,926,312)	(38.8%)▼	(39,650,488)	(40,687,950)	(2.5%)▼	(2,400,958)	(3,926,312)	
EBITDA	**419,558**	**(469,039)**	**(189.5%)▼**	**(7,337,914)**	**(6,528,245)**	**12.4%▲**	**419,558**	**(469,039)**	
Depreciation	(33,601)	(36,338)	(7.5%)▼	(328,126)	(332,130)	(1.2%)▼	(33,601)	(36,338)	
Step Rent	32,235	35,243	(8.5%)▼	229,384	406,605	(43.6%)▼	32,235	35,243	
EBIT	**418,191**	**(470,134)**	**(189.0%)▼**	**(7,436,656)**	**(6,453,770)**	**15.2%▲**	**418,191**	**(470,134)**	
633501 - Business Fees & Taxes	-	-	-	-	-	-	-	-	
690003 - Misc Income or Expense	-	-	-	(177)	-	-	-	-	
690004 - FX Transaction Gain or Loss	249,668	-	-	(15,979)	-	-	249,668	-	
695900 - Interunit interest income	-	-	-	-	-	-	-	-	
696000 - Interest Income	1,611	-	-	18,110	-	-	1,611	-	
MISC	**251,279**	-	-	**1,954**	-	-	**251,279**	-	
Net Profit	**669,470**	**(470,134)**	**(242.4%)▼**	**(7,434,702)**	**(6,453,770)**	**15.2%▲**	**669,470**	**(470,134)**	

Planning View example:

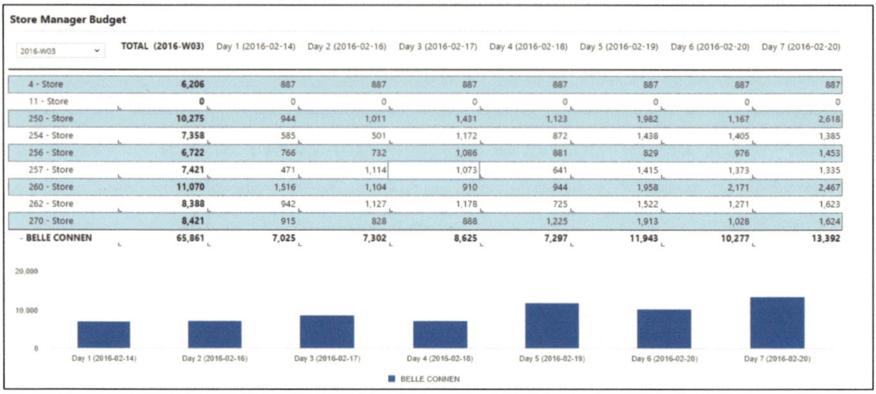

View Themes

Themes allow you to apply consistent colors, fonts, and other characteristics to your view. There are several standard themes that come with the default Jedox install, or you can also create your own corporate theme.

You can set a default theme at the user or group level, so every time a view is created the specified theme is automatically applied. The theme editor (part of the view menu in spreadsheet Designer on the web) allows you to change various parts of the view, including the title, subtitle, row headers, column headers and data cells. You can optionally lock or unlock cells. The theme editor can even conditionally format consolidated data (with limitations).

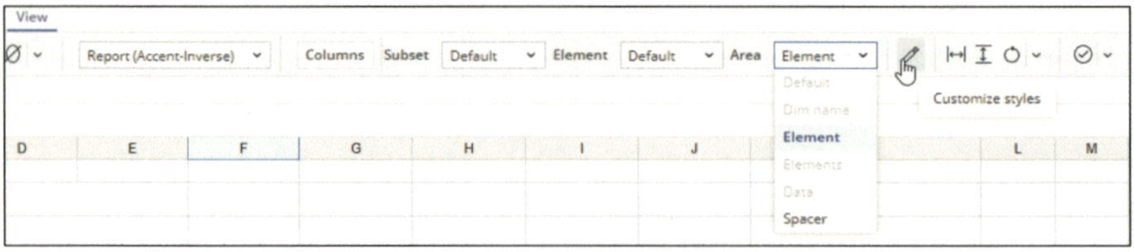

View Lists

Lists are a powerful, data-oriented feature that was introduced into the product recently. Lists are a component that define various combinations of elements from different dimensions. This enables the designer to create the exact columns required by the report, regardless of what dimension the data is coming from.

Without lists, using views are sub-optimal as you would only be able to use a single dimension to define rows or columns. View lists bring spreadsheet-like flexibility to the report definition. View functionality also includes the ability to use spreadsheet variables. This allows the designer to create combo boxes or list boxes for users to select the values for the view, which makes the view much more report-like.

The List Designer:

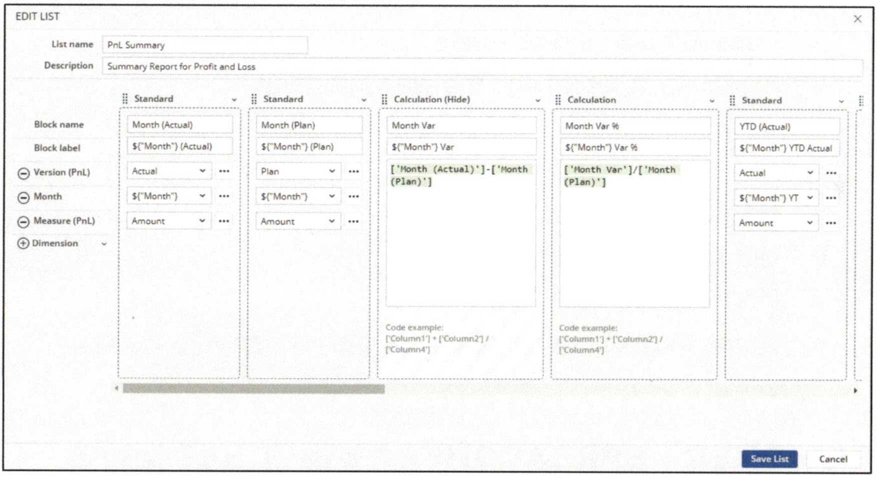

The same List in a View:

Oct (Actual)	Oct (Plan)	Oct Var %	Oct YTD Actual	Oct YTD Plan	Oct YTD Var %	FY (Actual)	FY (Plan)
5,079,655	5,798,579	(12.4%)▼	51,057,177	54,856,121	(6.9%)▼	5,079,655	5,798,579
(2,083,738)	(2,155,554)	(3.3%)▼	(22,283,428)	(22,829,526)	(2.4%)▼	(2,083,738)	(2,155,554)
1,458	(173,957)	(100.8%)▼	(543,010)	(1,645,684)	(67.0%)▼	1,458	(173,957)
(2,082,280)	**(2,329,512)**	**(10.6%)▼**	**(22,826,438)**	**(24,475,209)**	**(6.7%)▼**	**(2,082,280)**	**(2,329,512)**

If you are time constrained or performance is an issue, try to use views where possible in lieu of spreadsheets. It is a much faster experience for users.

Decision time – When is it best to use views as reports in your Model?

Pros:

- Fast
- Easy to build and maintain
- Available in Excel and the web interface

Cons:

- No access to data validation
- Not across multiple cubes
- Some cell formatting difficult
- Some form elements not available for use with views

DESIGNER - SPREADSHEET BUILT REPORTS

The Jedox spreadsheet designer is the workhorse of report building. This component allows you to build nearly any type of report in Jedox. If you can build it in Excel, you can build it in designer. It allows you to achieve pixel perfect corporate reports, planning templates, and management reports.

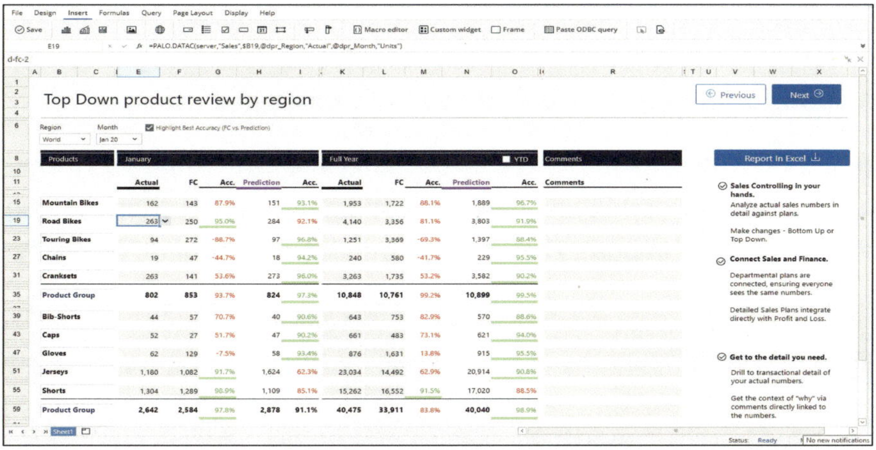

As well as building reports, Designer allows true applications to be built in Jedox by non-programmers. As well as in-cell formulas that replicate Excel, there is also a collection of Jedox formulas to pull data directly from the database, plus the standard view paste functionality. Instead of writing the formulas manually, there is a query toolbar which allows you to point and click to select elements and data directly from the database. In addition, there is a full suite of form objects (some familiar from Excel, some new) that provide unique capabilities.

Let's review some of the unique components of designer that makes it so special.

Dynaranges

Vertical and Horizontal Dynaranges allow for the creation of controlled, dynamic reports. These components allow the report builder to select an array of elements from the database that can be dynamic when clicked upon by the end user. For example, the report builder uses a vertical Dynarange with regions on rows. When the user clicks on All Regions, the report expands down to the next level. The same can be done on columns, using a horizontal

Dynarange, for example the expansion of months. Data within this component can be formatted using conditional formatting, making it a powerful tool for the construction of dynamic reports.

While the above example is also possible in views, when you report has a complex combination of data sources (e.g., accessing data across multiple cubes), Dynaranges provide the required flexibility.

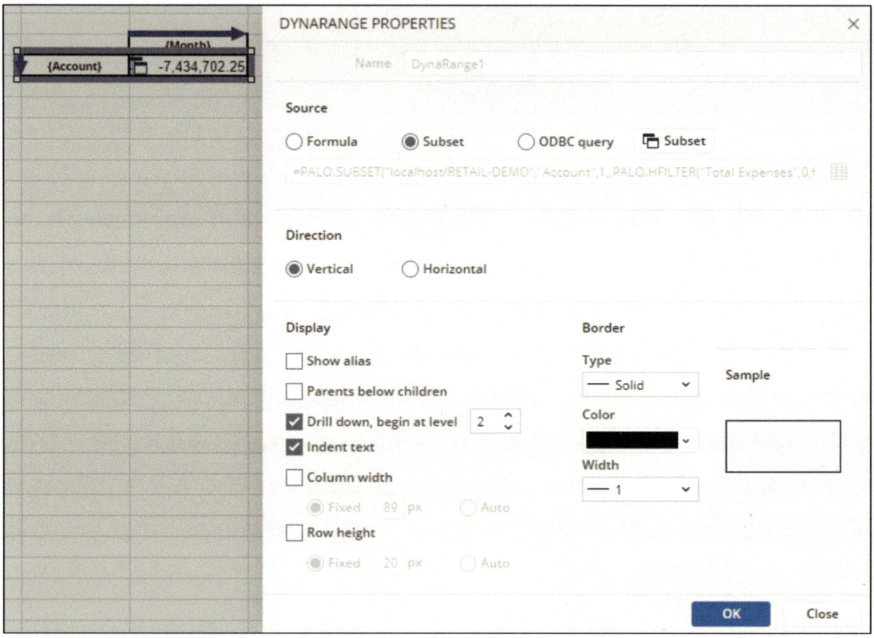

Actions

Actions allow common activities on a report to be built easily in a form/wizard approach, including the ability to create dialog boxes for the user. For example, in payroll/headcount planning, a user might want to add a new employee to the model for the upcoming year, load data from an Excel file and run an integration process.

Actions allow all of this to happen and can be driven by variables in the spreadsheet. They can be chained together and executed sequentially.

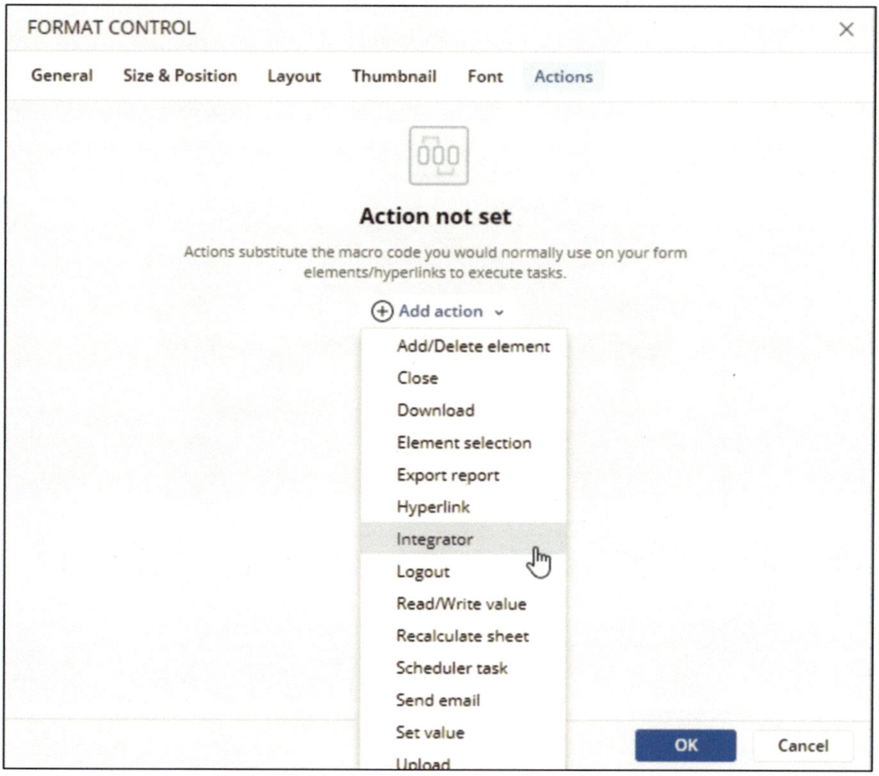

Workbook Events

There are three key workbook events that can be set in Jedox: when opened, when loaded, and when closed. Each event can have Actions assigned. This gives a lot of power to the report designer to set variables and other values on the report. The report can, for example, know the users cost center, location and workflow stage as the report is opened. All values can be prepopulated which makes the experience of entering the plan numbers much easier and seamless.

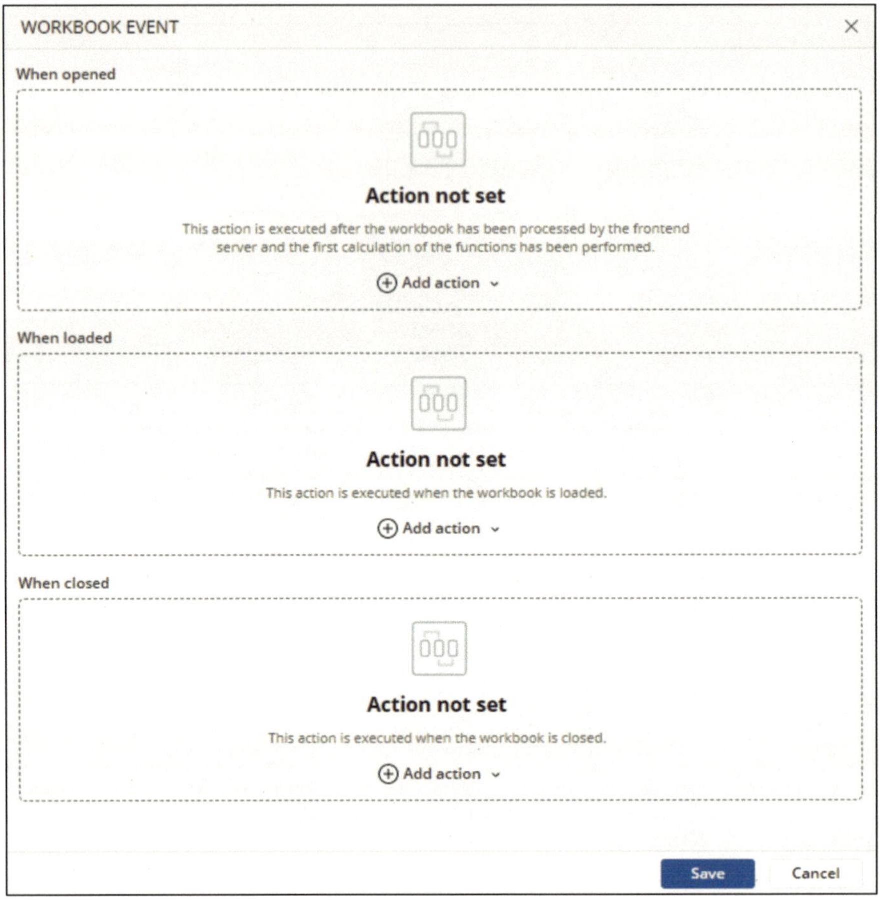

Framesets

Framesets are not exactly a worksheet component - they are a report type. Framesets allow multiple reports to be combined to create much more of an application feel. For example, there might be a common header you want to use on all reports, or a common side menu. These can be included in Framesets so to the user it appears like a singular, unified experience.

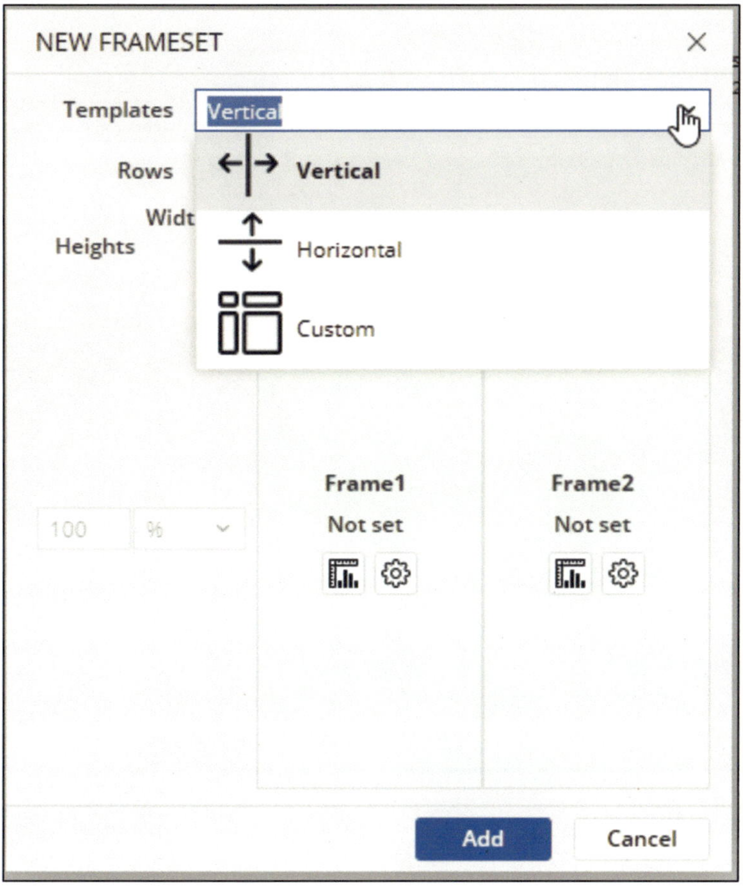

Custom Widgets

Widgets allow the report builder to include visualizations, reports, data, or website content not typically available in Jedox. Widgets allow the builder to enter a URL or HTML code. Variables can be passed in and out of widgets.

For example, there might be a specific chart not available in Jedox but in a third-party system that the report builder wants to include in a report. With a widget, this is possible to do.

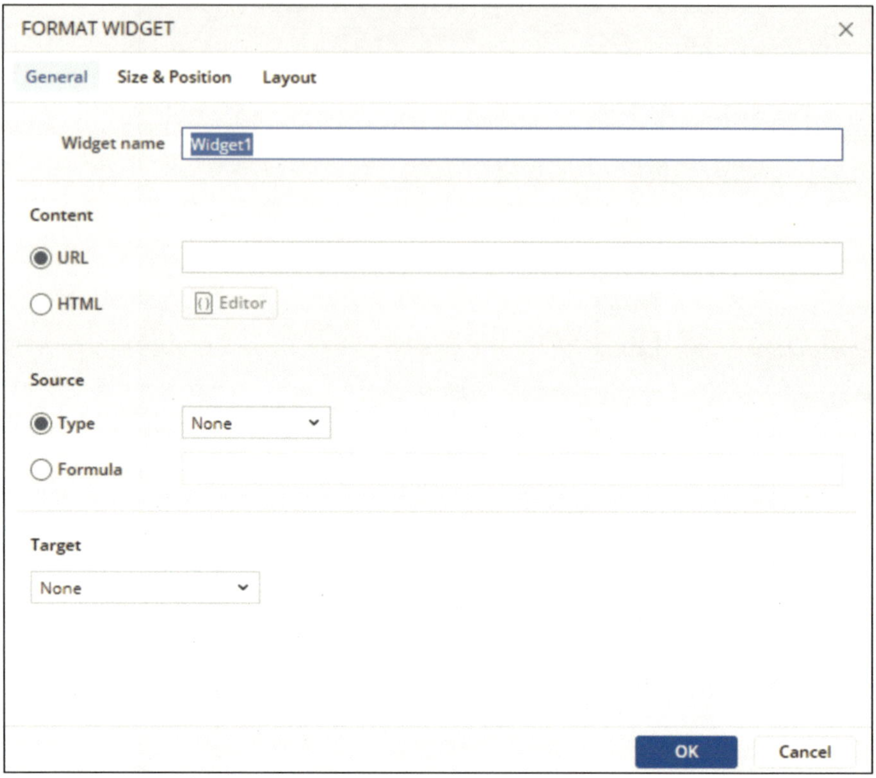

Macro Engine

Jedox has an inbuilt macro engine for scripting, automation, and custom functions. It is rarely used for standard functionality with the advent of Actions. All the methods and functions of the macro API are described on the knowledgebase. Macros can be executed via buttons on the spreadsheet, or as volatile custom functions in the spreadsheet.

```
function example()
{
   // get the active range on the current worksheet
   $range = activerange();
}
```

Jedox Hyperlinks

Jedox hyperlinks are the same as hyperlinks you might find in Excel, but with one important difference: you can specify variables, named ranges or cells to be passed from the source to the destination. This is a powerful addition to standard hyperlinks as it makes navigation between reports/templates seamless and more application-like. Values on the current report can be used as inputs on the next. You can also control the size and location of the target window. This gives additional possibilities for the creation of pop-up dialogs for the user.

Jedox hyperlinks are available in cells, form objects, charts, and images. On charts, a hyperlink can pass the value, category label or series to another worksheet or object as a variable, named range or selection.

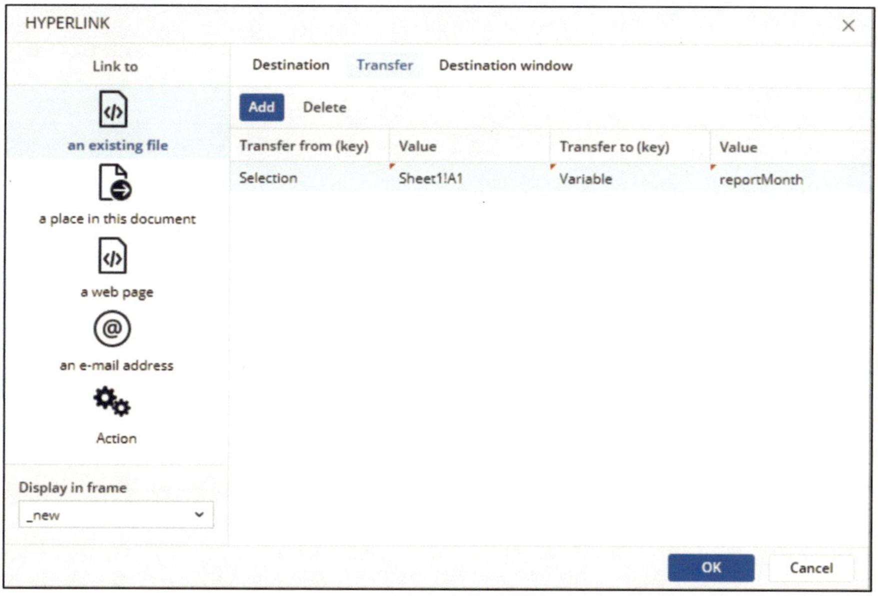

Each of these components are powerful additions to report building. Combined with standard spreadsheet functionality and data live from Jedox, almost anything is possible on a designer worksheet.

However, just because anything is nearly possible, it does not mean that all this functionality should be used all the time.

Remember, the golden rule for report design is ***simplicity***. The less Jedox needs to calculate on the spreadsheet, the faster the experience will be for the users (and the easier it will be to maintain). Try not to get carried away with the cool things that can be done in designer and focus on exactly what the deliverable is for your users, and how your design fulfills that requirement.

JEDOX CANVAS - BRINGING IT ALL TOGETHER

Jedox canvas is a report type that allows you to arrange and compose Jedox data in a flexible, scalable layout. It allows a visualization that goes over many pages deep and wide but retains smooth scrolling. It can be designed to fit many mobile resolutions and screen types and will automatically scale to the available screen resource.

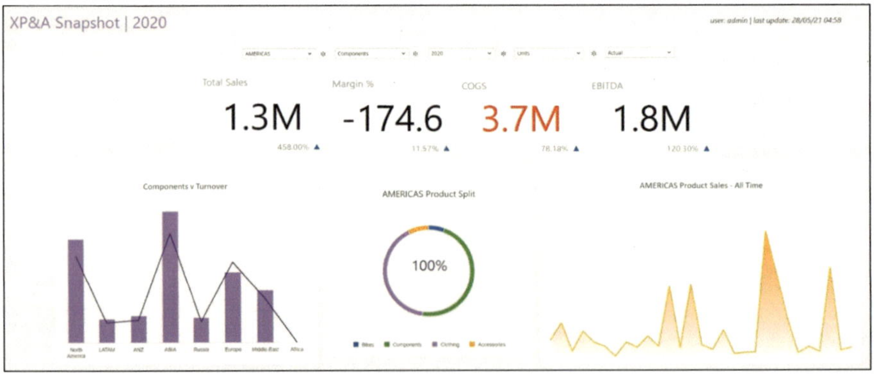

Currently, all content used in canvas must exist on a report, but the future roadmap will allow native components to be dragged directly. Canvas is not designed to replace Qlik, Power BI or Tableau - think of it more as a composition and dashboarding tool with writeback specific to Jedox components.

Content that can be used on Canvas includes individual form elements (charts, buttons, etc.), named ranges and areas defined in the report by framesets. You can also drag on a complete workbook tab. If you are utilizing content that includes variables (for example, a combo box) these variables will be shared with all content on the canvas that refers to the same variable. This allows you to have a very flexible canvas that can change dynamically based on user interaction.

Canvas allows the report builder to bring together content across many various reports into one, single unified view. This could be represented as a dashboard of visualizations, or even a "one-stop-shop" for planning. A single canvas could contain sales, opex and payroll planning templates so the user sees immediately the impact of their planning decisions.

REPORTS - PUBLISHING TO YOUR USERS

Once you have built your templates, reports, and canvases in designer, the next step is to make them available to your end users by publishing them to Reports. This process puts the report into "end user" view - it removes gridlines, toolbars, locks cells and the design components so they cannot be altered or seen by end users.

Publication Methods

Report tree

If you are going to use the tree control on the left in Reports to manage your reports, then it is a good idea to set up your folder structure beforehand. Once done, you can publish the report directly from the worksheet:

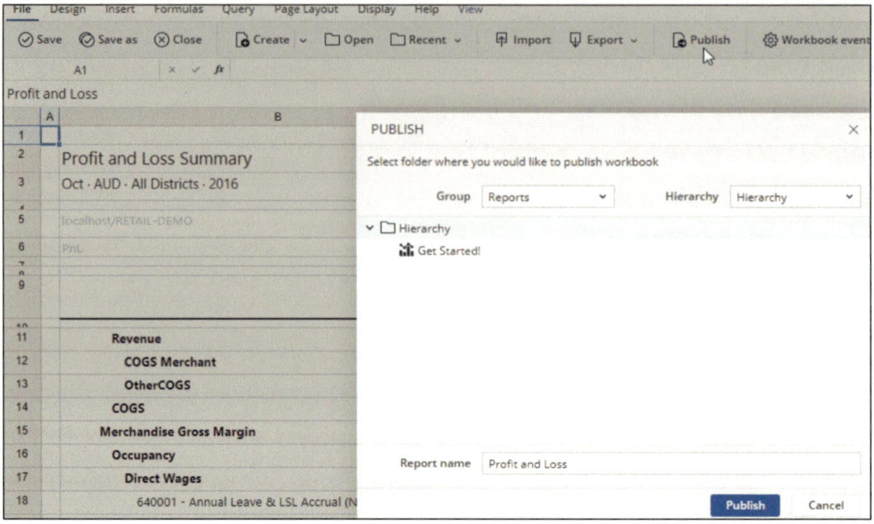

Alternatively, publishing can be achieved by dragging the file in Report design view to the applicable folder. By using folders and folder groups, you can apply security to hide and show them as required.

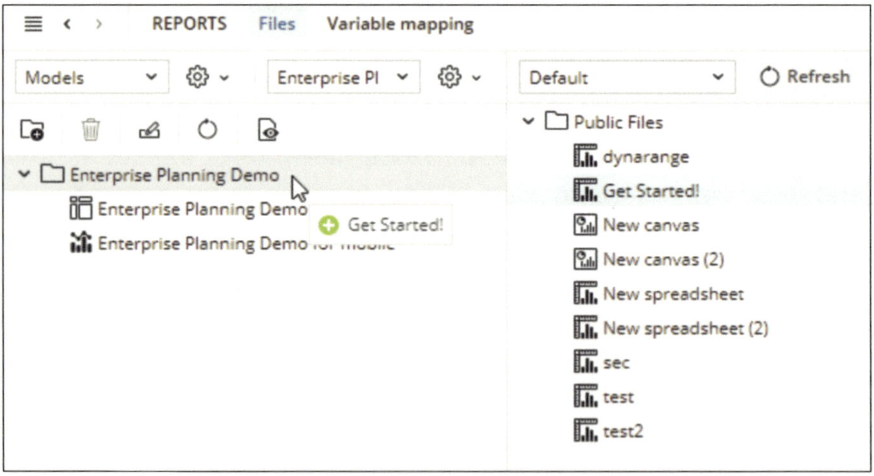

Customized Menu

Instead of utilizing the report tree in reports, you can build a custom main menu page (or pages) in designer with links to the appropriate reports. This can be as simple or sophisticated as you require. Links can be put in cells or on buttons. Report hyperlinks can be shown or hidden based on users' security level, and the whole menu structure can be driven by a dimension that is set up for this purpose.

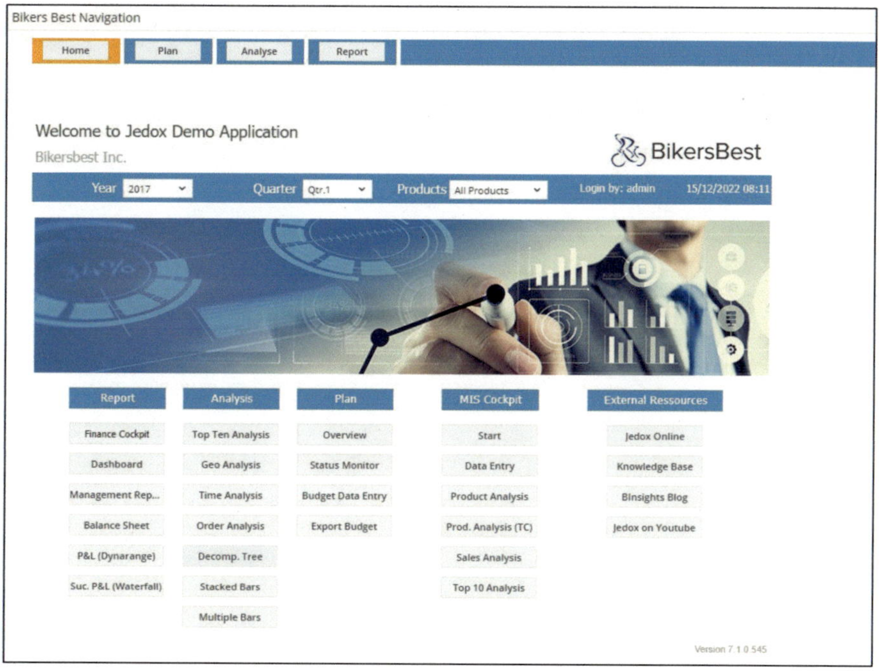

Embedding in Third Party Applications

It is not uncommon to embed a report or complete menu into a third-party application, like SharePoint, Power BI, or Salesforce. Published Jedox reports have a properties tab which enable you to apply certain settings. For example, typically you would want the link to open with that user's credentials so appropriate security and settings are applied. This means the user can directly enter and update their sales forecast inside Salesforce or Power BI with the advanced Jedox planning tools available. You can also choose to remove menus and frames from the report if required.

Chapter Five

Planning in Jedox

Planning is what Jedox does best. Jedox planning closely aligns with concepts in Excel, which means the learning curve of Jedox is generally lower for anyone with Excel experience. The multidimensional nature of Jedox makes it a clear fit for Excel - each data point in the Jedox database is represented by a single cell in Excel.

BASIC WRITEBACK

By typing over the top of a cell with a PALO.DATA style formula, the value is written into the database directly, and the formula reappears, with the last written data value shown. This enables changes to occur in real time across the whole model, as the newly changed values are available throughout the whole system, including all aggregations that use this value. If rules of other calculations are based on this cell, then further calculations are triggered.

For example, a new employee might be added into the payroll model with the associated salary and oncosts. If driven by rules, these cost values are automatically calculated into the employee cost lines in the Profit and Loss model.

This is the "basic" writeback functionality in Jedox. Attempting to write back at an aggregated level using this approach will give you an error:

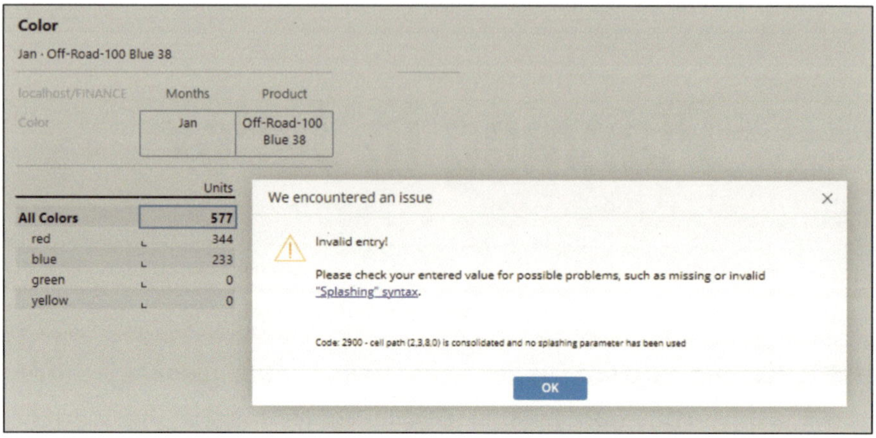

You can certainly writeback at a consolidated level, but to do that you need to use another type of writeback, called "Splashing".

SPLASHING

Splashing is where Jedox's true strength in planning lies, especially for reforecasting and adjusting a plan from the top down. Regardless of the model you have built, it allows you to shift huge values around in the underlying cube quickly. **Note, not all commands are described here**. For a comprehensive list, please see the Knowledgebase. For now, let's look at how the most frequent commands are typically used in Jedox planning processes.

PROPORTIONAL SPLASHING

Using the key character "#" allows data to be entered into a consolidated cell, proportionally spreading that value to all base cells. For example, entering #100 would proportionally spread that value across all base cells, so that the cell you entered would now be 100.

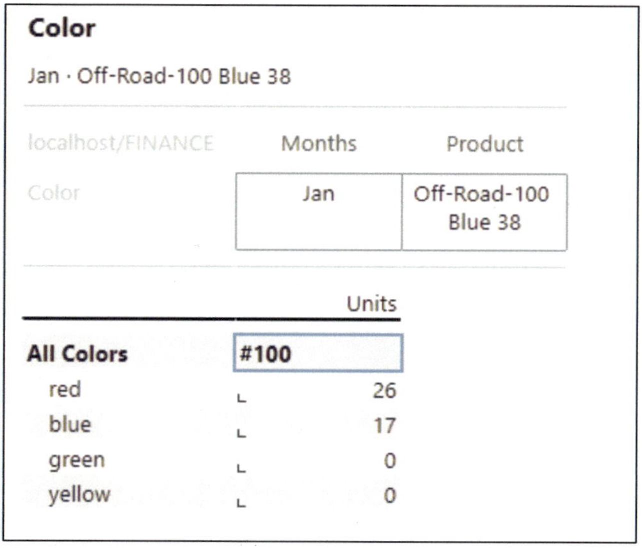

Entering ##100 would do the same, but it would **add** the 100 proportionally to all base elements.

Instead of whole numbers, you can also use percentages to change the value up or down: #10% - would increase the values proportionally to 10% of the original value:

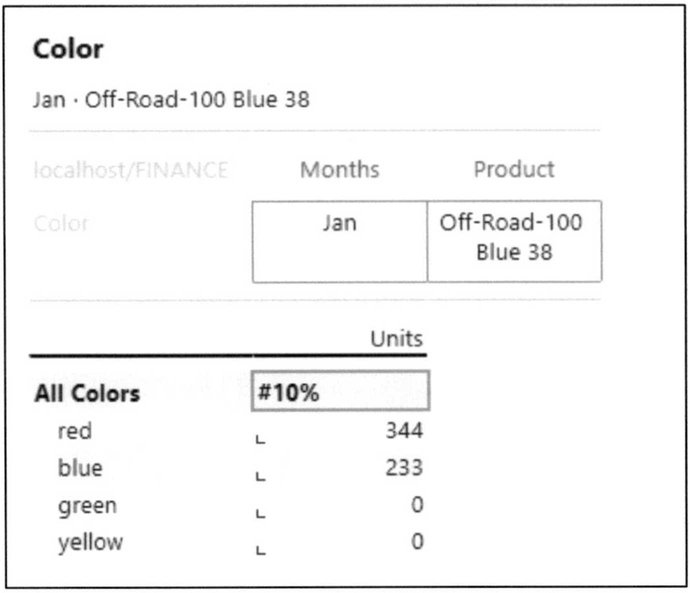

##-25% – would decrease values proportionally by 25%:

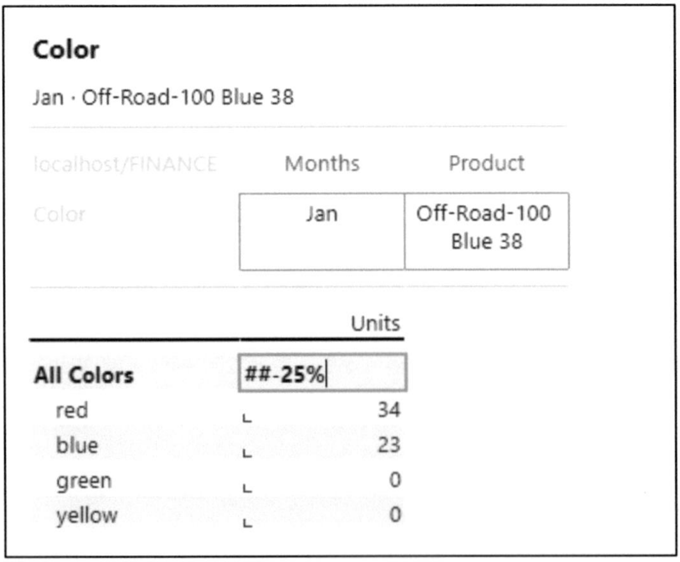

Proportional spreading using # in all its combinations is a fantastic top-down tool for management. It allows a management level view of planning and adjustment of numbers accordingly. For example: increase sales in division x by 15% over the next half year. This would change the values for all products, customers, locations for that division instantly.

COPY COMMAND

The copy splash command allows you to shift values from one area of a cube to another. It copies all base values released to the selected consolidated cell to the newly specified area, using the distribution of the source cell.

For example, copy the actual values from a product sold last year to the budget of a brand-new product releasing this year, as the sales profile should be similar.

In a cell pointing to Peripherals for 2022 for the Budget version, type:

copy Monitors;2021;Actual

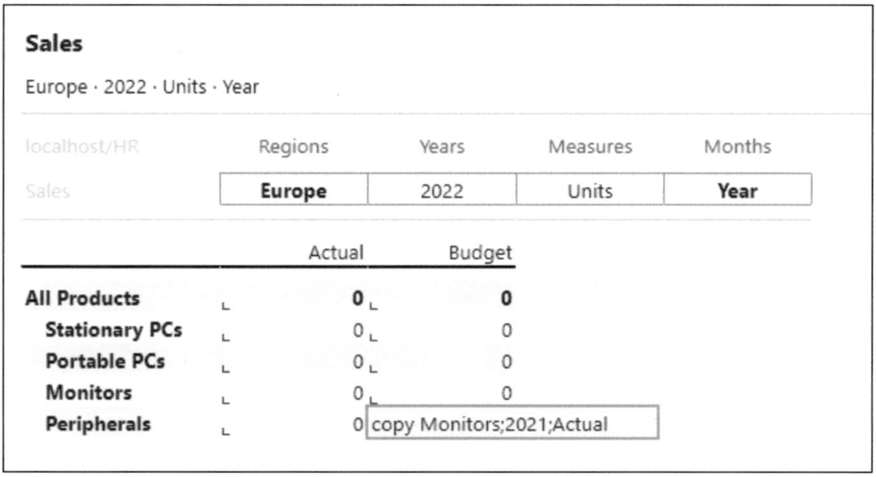

This would copy all base cells, proportions, and values from Monitors, 2021, Actual to Peripherals, 2022, Budget. This creates the sales profile for the new product. If required, it is possible to splash with the # command additional values or percentages across Peripherals to get a more realistic picture of the product sales. All of this is occurring at high level in the model, in seconds.

LIKE COMMAND

The like and copy commands are similar, however like allows you to specify a new number to splash across the new product at the same time.

In the above example instead of copying values and then performing a second command to # splash a new value in Peripherals, it can be done in one step by entering:

1200000 like Monitors;2021;Actual

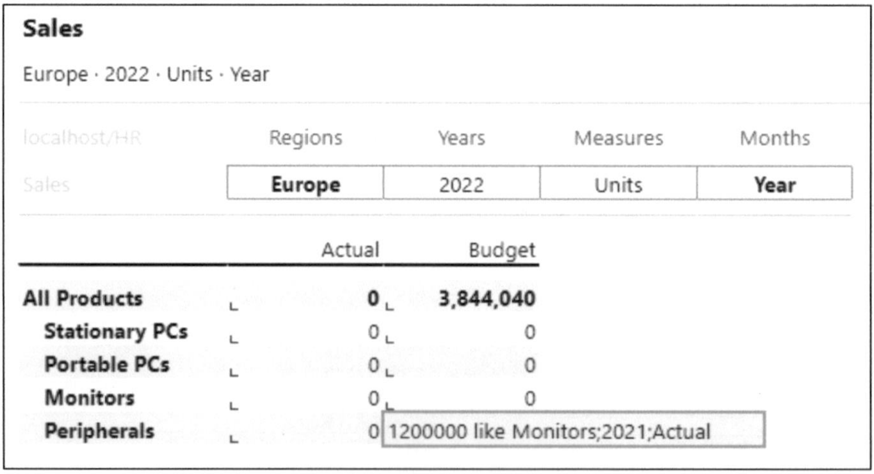

Now the new product receives the base value proportions that total the entered number, plus the new value across that distribution instantly.

TRANSFER

Sometimes, at a high level you might not want to change the overall planning number, but want to increase one area, while decreasing another by the same amount. The Transfer (or "To") splashing command allows you to do this in one step.

For example, you might want to decrease Peripherals by $1 Million and increase Monitors by the same amount to compensate.

By selecting the Peripherals, you can write:

1M to Monitors

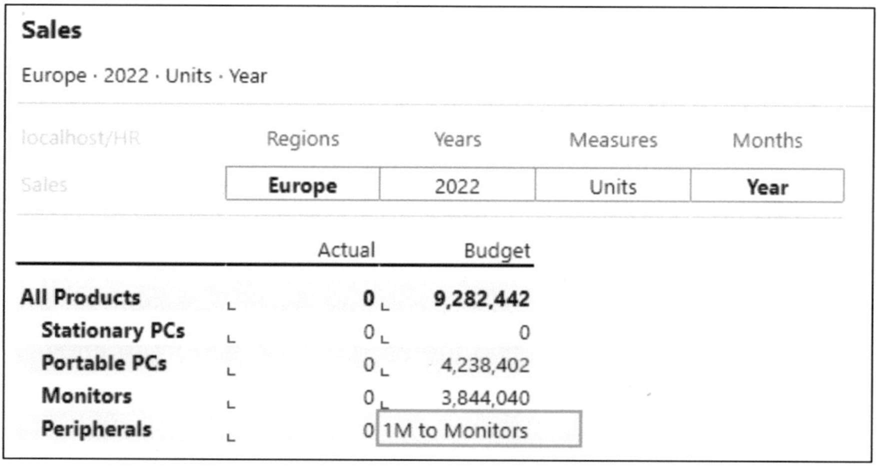

Result:

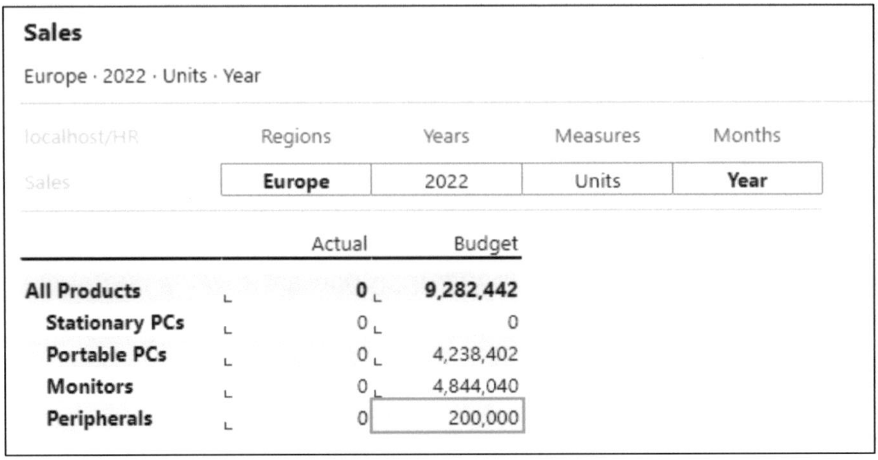

This increases Monitors by $1 Million and decreases Peripherals by the same amount but keeps the proportions of the cells within the Monitors amount. To use Peripherals proportions, or to transfer to an empty area, use the **FullFrom/FullTo** splashing command instead.

THE PLANNING ASSISTANT

The Planning assistant is available in the web interface and in Excel. Instead of using the short cut commands covered above, you can use the assistant in any cell that contains a PALO.DATA style formula. Some users find the wizard approach more intuitive than shortcut keys:

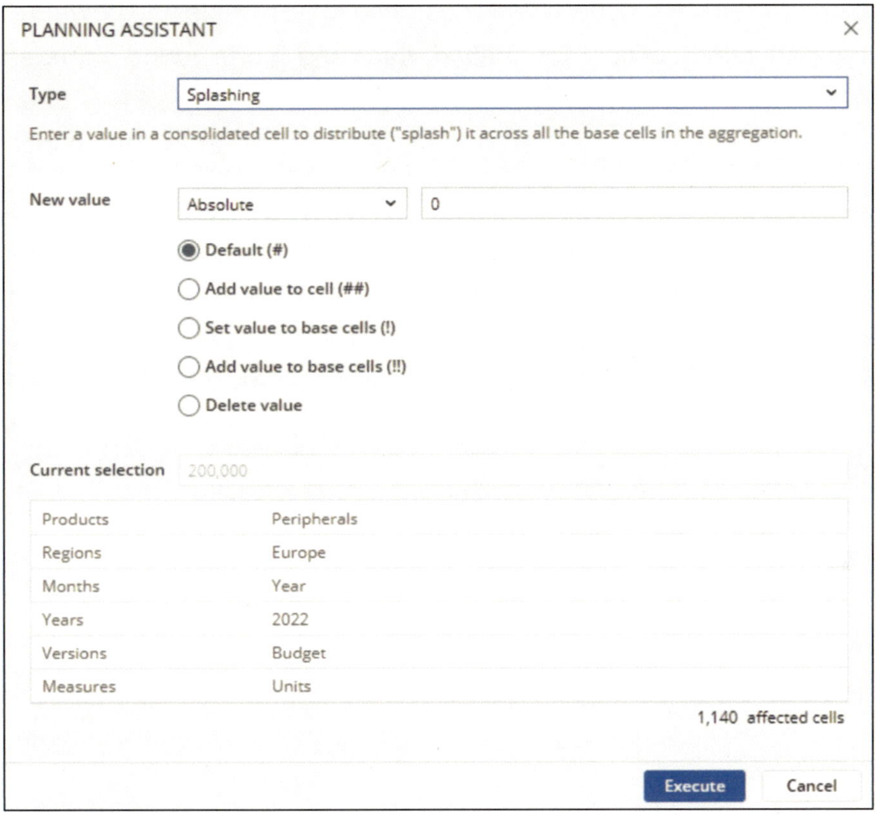

NUMBER SHORTCUTS

Instead of entering whole numbers or percentages, either at base or consolidated values, you can also enter number shortcuts. For example, you can enter:

1.5K instead of 1500
1.5M instead of 1500000
20% instead of 0.2

This saves time and accuracy, especially when entering values at an aggregated level in your model.

Jedox writeback functionality is designed to be flexible and powerful. It is particularly critical for second, third and fourth round changes in a corporate planning cycle where the bottom-up plan has been accomplished, and top-level changes need to be made in the model.

Traditionally, these "post phase 1" planning rounds are time consuming as each subsequent planning round is a top-level number but implemented bottom-up fashion. It is not uncommon for a first planning round to be complete in a week or two, but it might take months before the final round numbers are locked in and reconciled in the system. Jedox splashing technology shortcuts this from days/weeks to minutes/hours.

Chapter Six

Models: Prebuilt Solutions

As well as being a platform, Jedox also has the capability to deliver pre-built solutions. These are ones that can be installed directly onto the platform and used immediately. They include everything you require – the database, reports data entry templates and integration projects.

Models can be installed manually (from a filesystem) or automatically installed from the Jedox Marketplace.

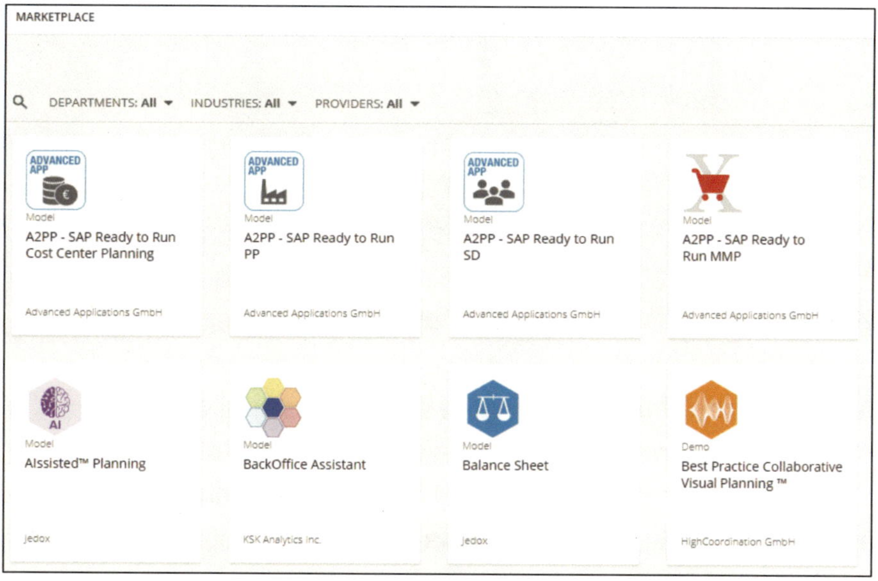

WHY USE MARKETPLACE MODELS?

Jedox is a rapid development platform. It allows you to design and implement a planning solution quickly that is very specific to your use case and organization. So why bother using a model? Well, there are some situations where the concept is very specific and complex, and the model can do a much better job than something you can build yourself.

For example, on the Marketplace, you will find the **Jedox Financial Consolidation model**. This solves complex financial consolidation between related entities in a corporate group. It contains similar functionality as pure-play financial consolidation software in the market, but with the advantage that it is all contained within the Jedox platform.

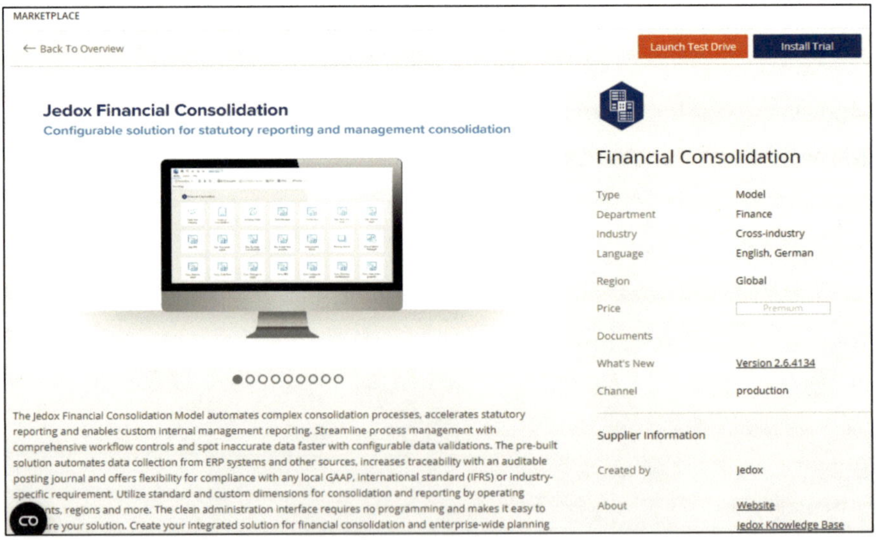

Another model that makes a lot of sense to use out of the box is the **AIssisted Planning model**. It takes advantage of the Jedox AI/ Machine Learning capabilities and can apply algorithms easily to an existing model that you might have in your system. It is wizard based, flexible and easily implemented with no data science knowledge. It is a unique model, as it provides additional technical functionality rather than a prepackaged process.

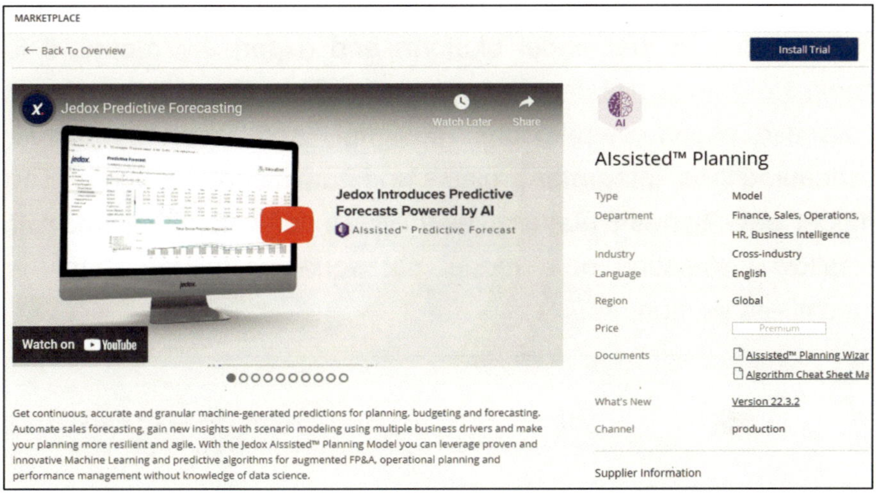

There are a lot of other excellent models in the Marketplace, including Profit and Loss, Balance Sheet and Cashflow models as well as domain-specific partner models. These are very good if you want not only a model but a process to follow for your reporting and planning. Some parts of the model are difficult to adapt if you want to change things (for example adding additional dimensions to the underlying cubes).

Matching the right type of pre-built model to the right situation is a great shortcut to instant productivity. Think carefully whether the prebuilt models fulfill your needs and will be right for you and your organization.

BUILDING YOUR OWN PACKAGED MODELS

Another interesting use case is to develop your own custom marketplace-style models for use in your own environment. When you develop in the model platform and export the model, it is completely encapsulated in a jdxp file. This file will contain all database objects, workbooks, designer, and report structures, scheduled jobs, integrator projects and administrator settings. The model development platform also allows for versioning and will disallow installation of a model not equal or greater than the incumbent version.

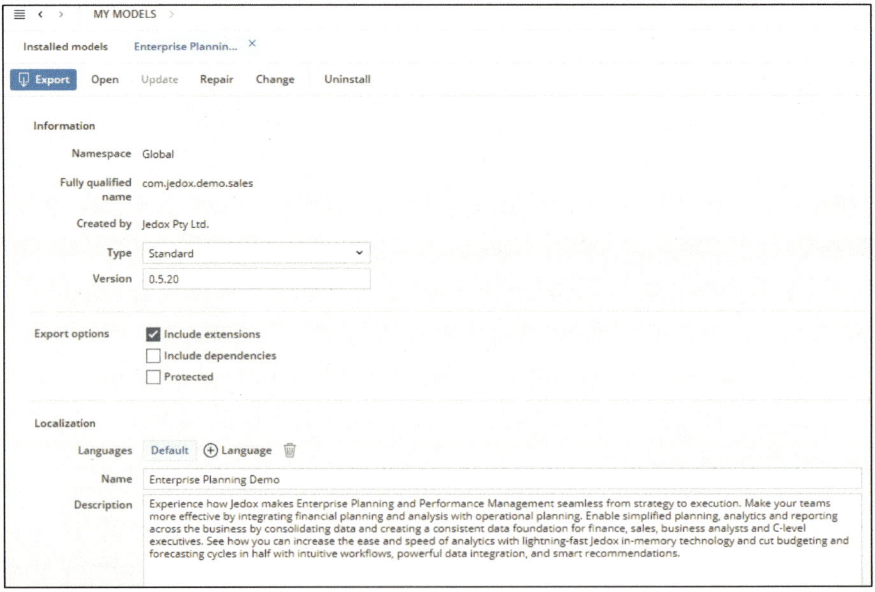

Packaged models, however, are not designed for change management specifically. There is no sophisticated change control, or the ability to update specific objects in the model (for example, a single dimension, report, integrator component). If your packaged model can exist and handle being fully replaced every time by the model updates, then this approach works. Otherwise, it is best to build your models directly on the platform, without using the prepackaged jdxp files.

Designing for Success

This section we will examine how we can use great design to make your model well structured, performant, and resilient.

> *"Design is not just what it looks like and feels like. Design is how it works."*
> *(Steve Jobs, co-founder of Apple, Inc.)*

In Jedox, there are many ways to solve business problems, but not all ways are appropriate all the time. You can deliver on all the requirements and still have a failed project because some aspect of design was missed.

Another reason a specific design phase is missed in so many projects is that as it is so fast and easy to model and build directly in Jedox. It seems initially cheaper to keep changing the model in Jedox until you get it right. But there are very good reasons why every project should have a specific design phase:

- Complexity of underlying data or business processes may require careful thought to ensure an optimal outcome
- User base size, network latency, geographical spread/time zones (that affects system up-time) and source system refresh times must be considered
- It will help you communicate much more effectively the outcome of the project to all stakeholders before you start
- Mandated corporate design standards need to be considered and incorporated into the project

It is imperative to include time (**normally one day to one month, depending on the scale of the project**) in your project plan to allow for a design process.

What does good design look like in a Jedox project? Regardless of the scope, a successful design will typically the following **key characteristics**:

- **Performance**
- **Scalability**
- **Aesthetics / ease of use**

Design Layers

Design can be looked at as layers in Jedox:

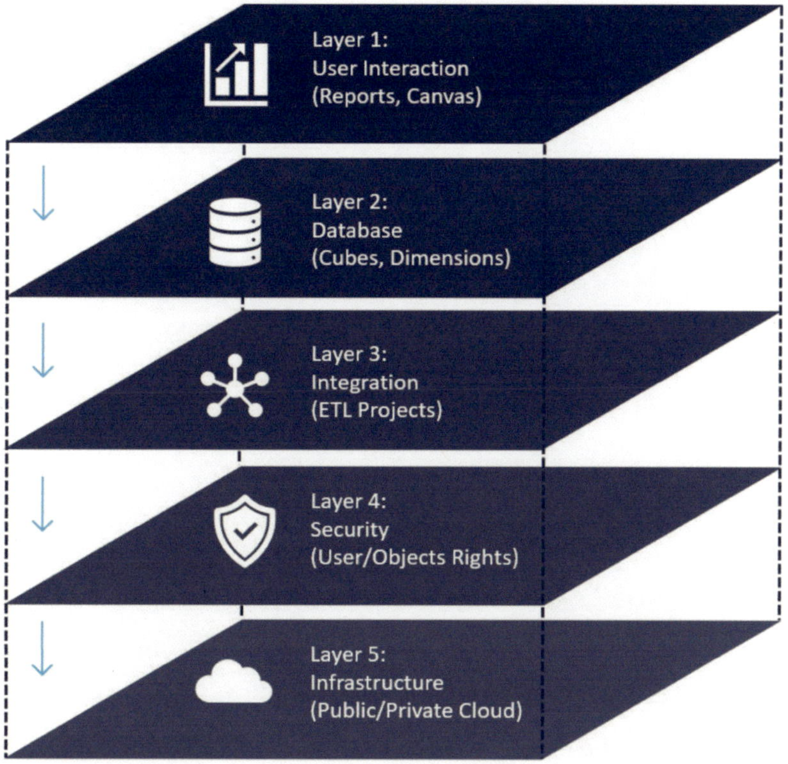

Each of these layers, and the interaction between them, can impact how your users experience performance in Jedox.

Remember, the design discussed here are suggestions, not law!

There may be very good reasons why you are running your Jedox instance in a particular way, that might go against the suggestions outlined below.

Chapter Seven

Design for Structure and Consistency: Build it Well

> ***"There is only one thing worse than Excel hell - and that's multidimensional hell".***
> *(Anonymous Financial Controller)*

Left without a common structure discipline, multidimensional models can grow like mushrooms and can become a very, very complex system over time. Designing for structure and consistency means looking at your model to ensure your naming conventions, descriptions, dimension orders, attributes, integrator projects, and report construction are all aligned and apply the same standard everywhere. By applying this discipline, it will be much easier to understand for new people (if existing personnel leave the project, for example) and make the existing model easier to extend in the future.

The key question that you should ask yourself is: ***in the absence of user manuals, system diagrams, and handover docs, would a new person in charge of this system be able to understand what I built just by looking at the model?***

The answer you should be striving for is a resounding "Yes". Even if you have diligently created every document and schema for a new administrator/developer, chances are that they will struggle to read or understand what you have written, or not bother to read it at all. Fortunately, by building a model with consistency and good structure, you can make your system very understandable, even without external documentation.

To recap, some definitions from the Glossary:

Structure: Means how objects are laid out or presented in Jedox.

Consistency: Means the definition of a "common language" that defines objects in Jedox and how it is used throughout the model.

Let's look at structure and consistency design through the layer lens.

STRUCTURE LAYER 1: USER INTERACTION (REPORTS, CANVAS)

To make your reports easily digestible for users, consistency is key. This applies right down to the naming conventions you have on the report and what you call the reports in the report hierarchy in the Reports section. You should pick a naming convention that clearly identifies the contents of the report. Here are some ideas to keep the consistency throughout your reports:

- **Use a Global workbook.** Jedox allows a user to tag a workbook as a "Resource". Once you have added that workbook as a resource in another workbook, you can then refer to it in spreadsheet formulas in other workbooks, just like linked spreadsheets in Excel. This enables you to store key values, names, and data centrally and use it dynamically in all connected workbooks. It will also reference any custom style sheets you have created in the global workbook to use across your reports.
- **Ensure reused objects are consistent in both function and meaning across reports**. Named ranges, variables and other common values that are used in your reports should have identical names across all reports.
- **Clear report naming convention.** Ensure you have a sensible naming convention across your report suite that helps users identify the contents of the report without opening it. Jedox allows you to rename reports when you publish them, so it gives you another opportunity to refine your consistency. Be very clear on how you are going to name and describe things. For example, you could specify that report names will always end with "Report" or "Plan" and only be a maximum of 50 characters long.
- **Use Report Description fields.** In Designer, when you right click on a report, you can view the report's properties. Here you can set a description that is saved with the report. The description field lets you provide a detailed explanation of the report. Make sure that even with descriptions, you create some rules around it. For example, you may decide all descriptions will only be in one language (e.g., English). These description fields will probably not be used by end

users but will be a big help for new system admins in the future.

- **Keep the use of Spreadsheet macros to a minimum.** Macros were frequently added to Jedox reports by advanced users. Not anymore. With the addition of Actions, 80-90% of all macro scripts can be replaced with a dialog driven Action, which is clearer and easier to understand by any level of user. If you must use a macro, ensure that the code is clearly documented with code comments so it can be clearly understood in the future.
- **Conditional Formatting and Style Sheets.** To keep a consistent look and feel, be sure to set up your custom style sheets in the global workbook and share with each report. This ensures that in the future, if there are changes to the design styles, the style can be changed in one place and propagated everywhere.

STRUCTURE LAYER 2: DATABASES (CUBES, DIMENSIONS, LISTS)

If you only have time to focus on one section of Structure and Consistency, this is the most important. Well structured, well-named and consistent cubes, dimensions and databases make a huge difference in understandability and data transfer through your model. A Jedox environment can become complex and large over time.

This can sometimes lead to objects that are similar or identical being called different things. For example, you could have a time dimension called "Time" that contains month elements in one database, and the exact same elements in another dimension called "Date" held in another dimension. This is bad design. Not only is it confusing for new users looking at them, trying to find the difference between them, it is a technically bad design decision as well. You lose out on a lot of optimizations.

Now imagine you have 30+ databases, and the standard month time dimension is called something slightly different in each. As you can see, it gets complex quickly. Here are some thoughts on naming conventions in Jedox for database objects:

Resist the keyboard. When you start a Jedox project, your design process should begin away from the keyboard. Start on a whiteboard or with pen and paper. Why is this important? It allows you to focus on the business problem you are trying to solve, and not get distracted by the technology you will ultimately use to solve that problem.

Your design process for structuring a model should start with all the known information about what you want to achieve and laying that out as a visual that can then be used as a design map.

For example, you need to build a Sales model that feeds its data into a Finance cube:

JEDOX SALES CUBE DESIGN

Database: Sales

Sales Cube Name Monthly Sales

Description holds monthly sales data

Dimensions: Version (holds Actuals, Budgets, Forecasts)

Month (Group by quarters, year. Group by YTG and YTD)

Customer (Attributes: Address, rating, main sales team)

Product (Group by Subcategory, Category, Division. Attribute: Product Cost, Brand Name)

Location (Group by Area, State, Country. Attribute: Sales Manager, Cost Centre)

Sales Measures (Sale Value, Qty, Margin, Margin %, Sales Comments)

Integration? Load data from Salesforce nightly.

Planning Calculations Calculate Margin, Margin % for Planning Data.

Data Path through model Sales Cube -> GL cube

Data Path Mapping to GL Cube : Version → Version; Month → Month; Location → Cost Centre (map using the Cost Center attribute of the Location Dimension); Sales Measures → Account (Sales Value → "60667 Sales Revenue"); Margin → "60668 Sales Margin"

NON MAPPING DIMENSIONS: Customer → ALL CUSTOMERS; Products → ALL PRODUCTS

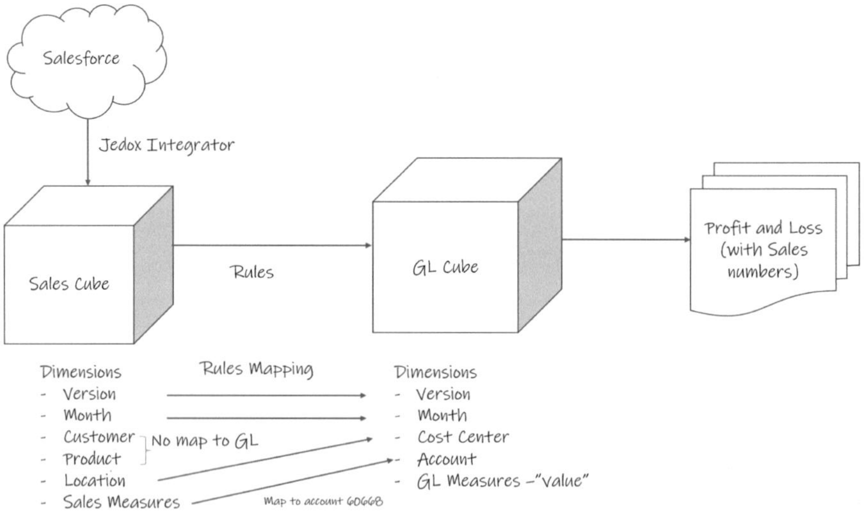

This sketches how the data will flow through the model. Visually documenting your design will help you understand what the final system will look like, without getting caught up in technical concepts like Jedox Modeler. Armed with this knowledge you can then request further feedback from the key stakeholders, until there is a clear consensus on how the model should work.

A final version of the design sketch should be kept as part of the record of the project's design decisions. Once you have an agreed upon final design, it is time to start constructing it in Jedox.

Firstly, let's look at how you should name and organize objects in Jedox.

Databases

Databases are great for collecting related objects together. Try to build each of your models so that each is contained within a single database. This helps segment your models clearly and makes the application of security and maintenance much easier. Here are some examples:

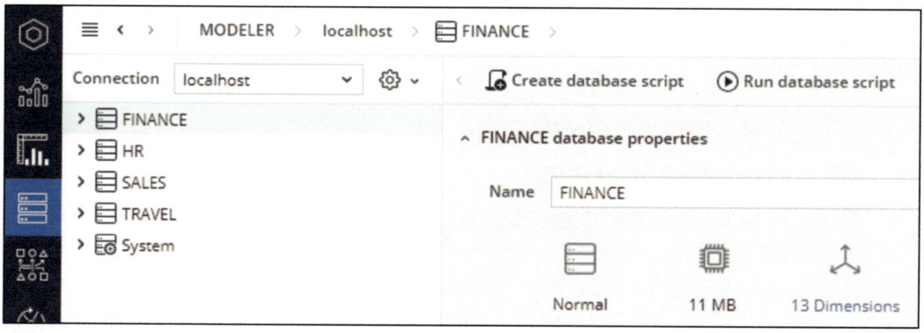

FINANCE - holds the profit and loss, balance sheet and cash flow model
HR - holds human resources and payroll data
SALES - holds sales relates cubes
TRAVEL - holds travel planning data

The names are short and understandable. Note that with database names, you can only use alpha-numeric characters, underscores _ and dashes -. Spaces are not allowed. For more information, please see the knowledge base section **Naming Restrictions for Jedox Objects**:

https://knowledgebase.jedox.com/jedox/modeling/object-naming-restrictions.htm

Cubes

Ensure the name of the cube is succinct, and clearly represents the data in the model. The advantage with cubes is that you have a cube description property. Here you can document in more detail about what information the cube holds.

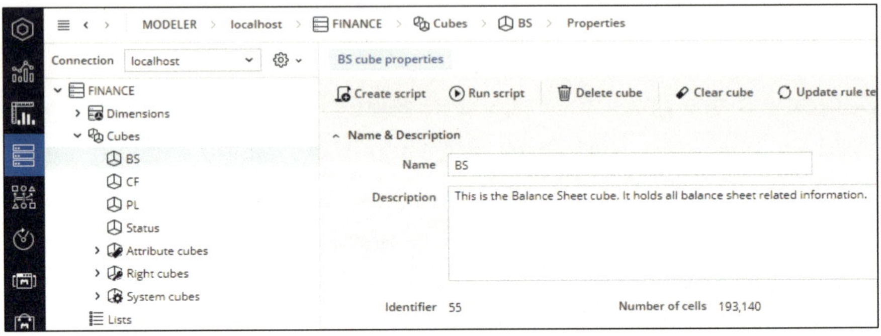

For example:

BS - The balance sheet cube
CF - The cash flow cube
PL - The profit and loss cube

When creating a cube for the first time, create an order of common dimensions, and stick to that order, regardless of which database they reside in. For example, you might have these common dimensions across most of your cubes:

- Version
- Time
- Cost Center
- Measure

It is good practice to end your cube with the measure dimension. You then insert in any specific dimensions above the measure dimension. This will ensure that dimension order is always the same for those common dimensions, regardless of the cube or database. This makes integration, rules, report building and views much easier to use and understand.

Note, Jedox does allow you to re-order the cubes, add and remove dimensions. The Jedox spreadsheet and rule formulas allow you to specify the dimension elements in any order you want (along with the dimension name). But it is always best to design it the right way up front.

Dimensions

Ensure the name is consistent across databases. If you have a product dimension in one database, and you have an identical product dimension in another database, make sure the name is identical. If the dimension is similar but different make sure you specify that in the name. For example, if a dimension holds only the group names of products, make sure you clearly name it “Product Group”.

There has been a debate amongst Jedox consultants for a long time whether a dimension name should be called a plural or singular name (for example, should the product dimension be called “Product” or “Products”?). No matter which method you prefer, ensure it is consistently applied.

An additional point should be made about dimensions that hold the value definition, or measure dimensions. These dimensions should be clearly labeled as “Measure” and should be placed consistently in the same place in all cubes. For example, it is common for the Measure dimension to be the last dimension in the cube. The measure dimension should typically reflect the cube name or the contents of the cube to which it is attached, for example: Sales Measures, Travel Measures, HR Measures. Keeping the measure dimensions separate means each one is small, and it is clear to the user that they all are relevant to the cube. Of course, measure dimensions can be shared across cubes like any other dimension. In this case, think of what those cubes have in common, or simply call it “Measures”.

Set the dimension properties as well, especially the default elements and dimension description. The description also allows you to describe in more detail the reason for its existence and how it is used.

Virtual Dimensions

Virtual dimensions are attributes from a dimension and are created dynamically by the database. They consist of 3 levels: the top-level node, values that represent the attribute “value”, and below that the elements that the attribute represents. The three-level structure is fixed, and you can only manipulate the dimension by changing attribute values. The big advantage is that the system takes care of the virtual dimension for you. In those situations where the levels and structure of a virtual dimension are acceptable to the model, you should use them to reduce maintenance and let the system maintain consistency for you.

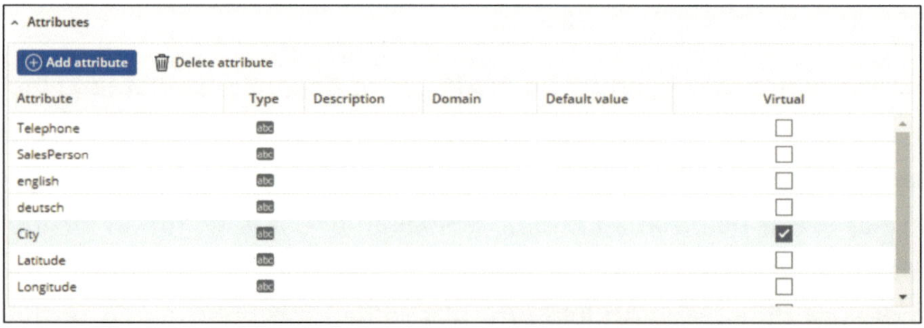

Elements

Element Codes and Renaming

Most of the time, the element naming conventions will be out of your control, especially if you are importing from a 3rd party system, like SAP, Microsoft Dynamics, or Salesforce. What you should do, however, is ensure that the actual element identifier in Jedox is represented by a code or number. You can then hold the name in an attribute and show that on the reports if required. This will ensure that if the description ever changes, you will not be faced with renaming the element (which is possible in Jedox) and reviewing your reports for the old name. Typically, IDs will not change in the source system, and it ensures you have a close alignment with the source data in your Jedox model.

Top Level Nodes

Usually, you will want a node at the top level in the dimension that you will need to create yourself. The standard convention is calling that element "All <Dimension name in plural>" or "Total <Dimension name in plural>":

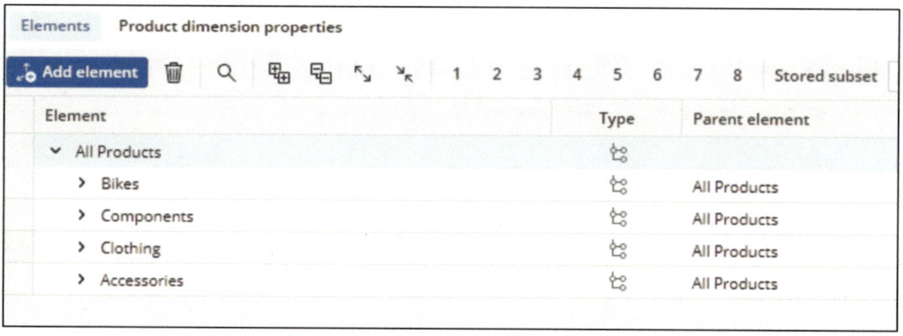

You can certainly come up with your own names - just make sure they are consistent.

Measure Elements

Most Jedox cubes include a measure dimension. This holds elements that represent a particular value. For example, a measures dimension could contain the following elements:

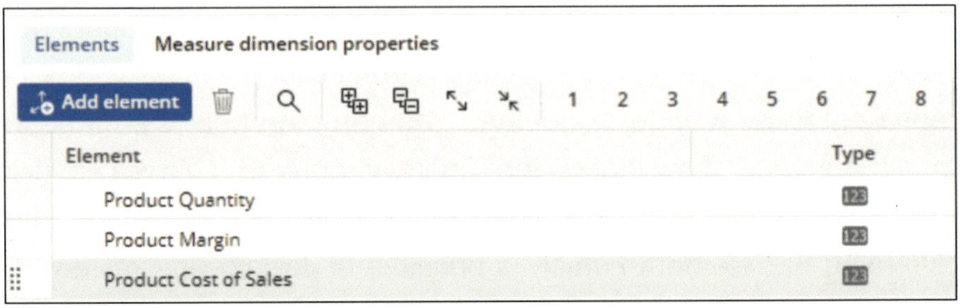

Like dimension names, it is crucial that whatever naming convention is followed, it is applied consistently everywhere. For example, don't call an element Product Quantity in one measure dimension, and Product QTY in another. This is just confusing. Another example is the reverse. If you have similar but different measure elements in different dimensions, make sure you clearly differentiate them. For example:

1st Measure dimension: "Product Cost of Sales"

2nd Measure dimension: "Product Cost of Sales" - but this one includes shrinkage. This is an incorrect name. For the 2nd measure dimension, you could either call the element:

"Product Cost of Sales incl. Shrinkage"

OR create an additional element representing shrinkage, and add both together:

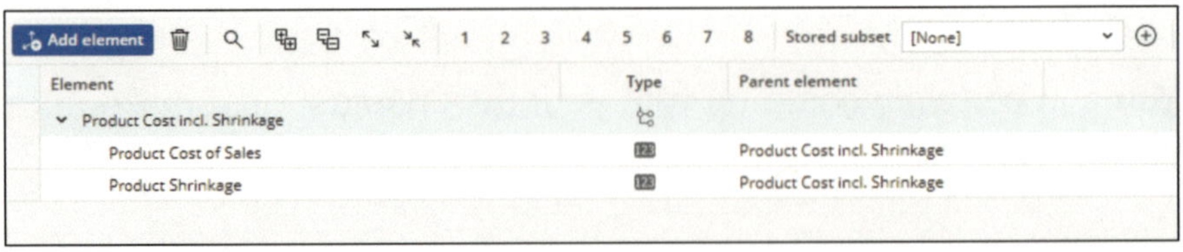

Lists

Lists are powerful objects to use in views.

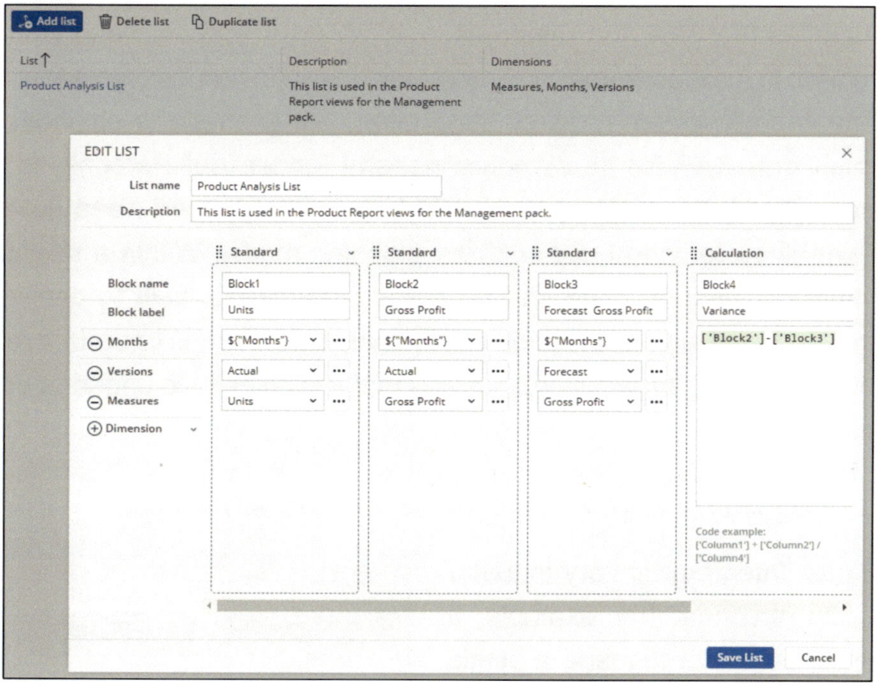

As they are essentially metadata objects, the main thing that you should be concerned about is the naming convention you use and a clear description to explain what it is and how it will be used.

The impact of a well-structured database is long-term. It makes management, development, and usability much easier and straightforward, as well as being self-documenting. It also ensures that the model can be easily extended and built on later without additional work to restructure the database.

STRUCTURE LAYER 3: INTEGRATION

Jedox Integrator is often the beating heart of any Jedox model. It supplies the data and metadata typically on a schedule into your model. In most corporate environments, it is common that more than one source system requires integration. For example, a business might run SAP for its standard general ledger, but Personio for payroll and Salesforce for sales. It is possible you will need data from all systems to build out your planning model. Within a single integrator project, there can be many components built to derive the final load step into a cube or dimension. Again, it is important to build your ETL project in such a way that it is as easy to understand as possible.

Taking Advantage of Integrator's Modular Design

Jedox Integrator is very modular in its structure. For example, it enables you to manage a single connection for multiple extracts and transformations within a project.

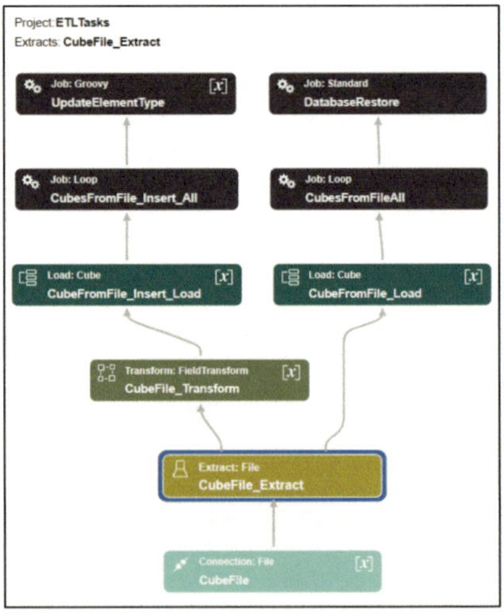

This means you can reduce the number of ETL components simply by sharing common components across your project. You can also use variables to determine component names and change the variable dynamically to switch components.

Keeping Connections Global

Inside Connections, it is possible to set up global connections to various source systems. This means that, regardless of the number of projects, you have a single point to manage the connection. This keeps the management of connections for Integrator simple and consistent.

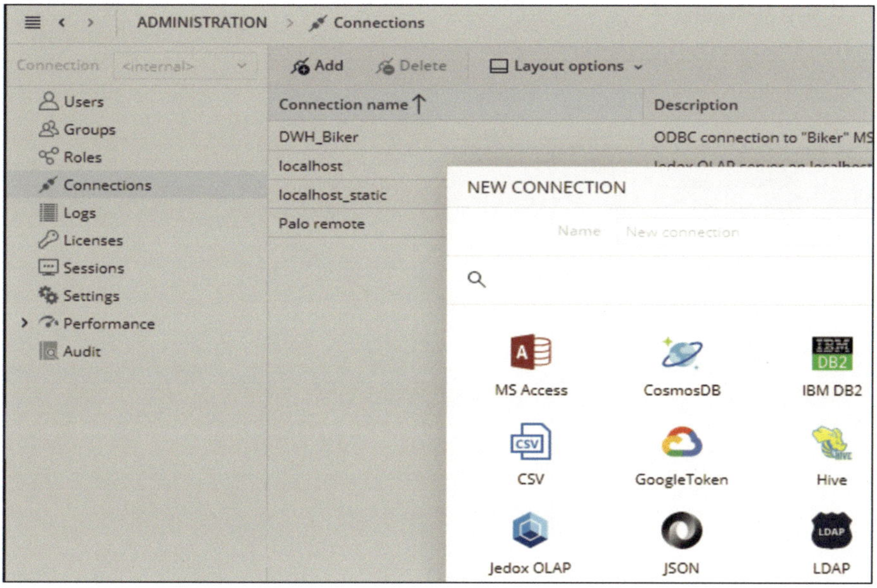

Naming Components

Integrator has no built-in name conventions, so it is a good idea to create your own. By looking at the component name, you instantly should be able to understand exactly what it is. The project name itself should be self-explanatory. Use the Project description field to provide an overview of what is happening in the project. NOTE: it is important to emphasize that there is no "perfect" way to name components – your naming convention must make sense for you and your organization. Here are some examples of commonly used conventions:

Example 1 - Simple Naming Convention

This is a very simple naming convention that describes clearly what each component does. This method works well when you have a small integrator project with only a few steps. The first letter (e.g., “L”) describes the type of component (e.g., Load, Job, Transform) and the rest of the name describes either the function or the cube you are loading to:

Sales data for 2022 from the data warehouse

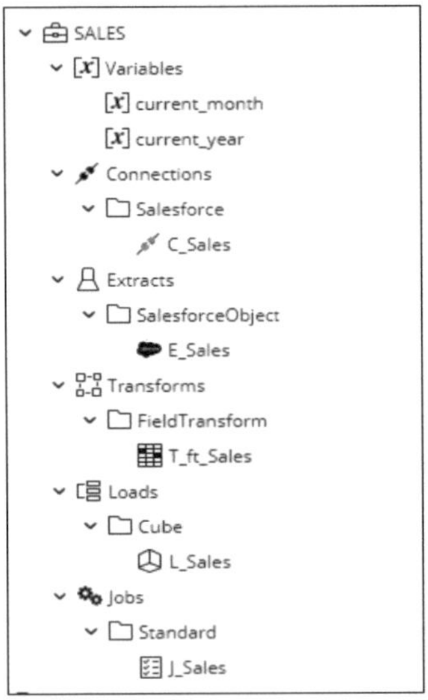

Example 2 – Medium Strength Naming Convention

This method works very well for a scaling model as it is easy to see immediately what each component does. The naming is short, and the prefix of the name describes its function group. This naming method works effectively when the project and process steps are quite linear – that is, there are clear before and after steps for each component. The example below is from the Jedox AIssisted model:

Example 3 – High Strength Naming Convention

This strong naming convention is particularly effective for very large projects where there are hundreds (or thousands) of individual components. Remember, scale equals complexity. When a project starts approaching this size, firstly consider whether keeping a simpler naming convention but separating out components into small sub-projects is more effective for readability and traceability. This naming convention is complex but has the advantage that individual components are easier to identify amongst hundreds and the name clearly describes what it will be used for. Of course, the Jedox tree view in Integrator shows these relationships as well; but in complex environments it is far easier to see the relationships if they have been documented in the names of the components:

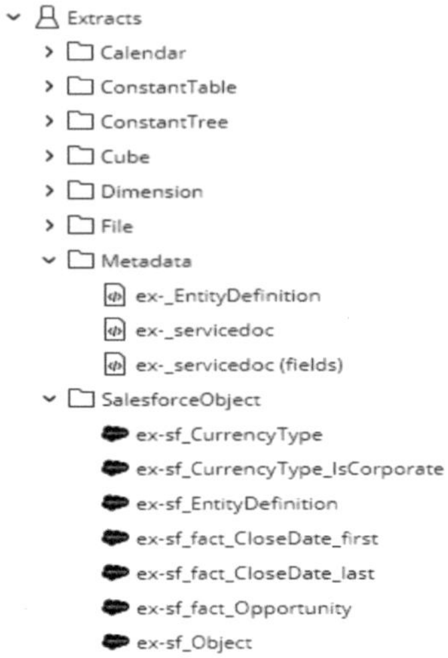

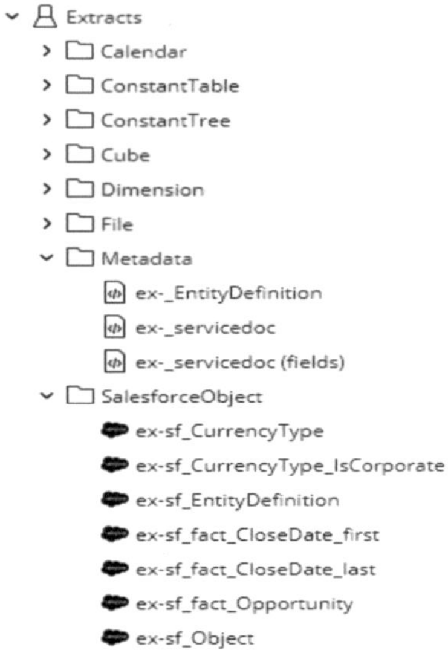

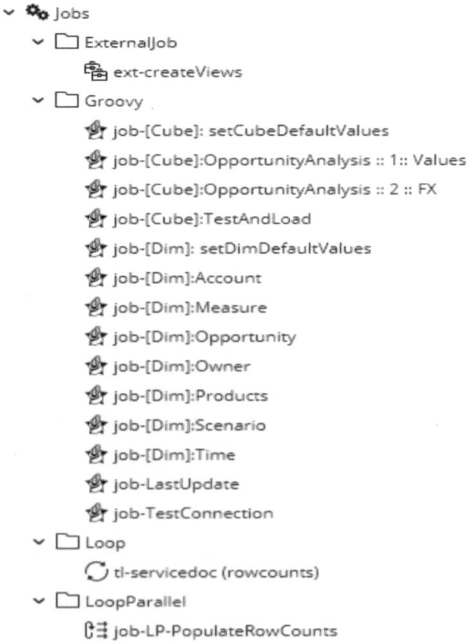

Explaining the naming convention above:

Extracts start with ex-<source>. The square brackets [] show the destination where the extract will be used. If the extract is going to be used in multiple areas, skip this part. The colon: is the separator, to the right on this is the explanation of what the data is.

Transforms start with tf-<2 or 3 code for the name of the transform> . For example, Field transform is TF, Table Union is TU, Tree FH is FTH, etc.

Loads and Jobs simply start with the word load- or job-. This is helpful, especially when running jobs or loads externally to differentiate what is a or a load. Additional prefixes can be added to differentiate job types (e.g. Parallel jobs, Loop, Loop Parallel, Switch, etc.) The name numbering pattern (e.g. :: 1 ::, :: 2 ::, etc.) in loads and jobs is used where the target is the same, and describe the order in which the component should be processed.

Component Descriptions

Every object in Integrator including the project, component and functions have a description field. It enables you to describe clearly what the component is doing.

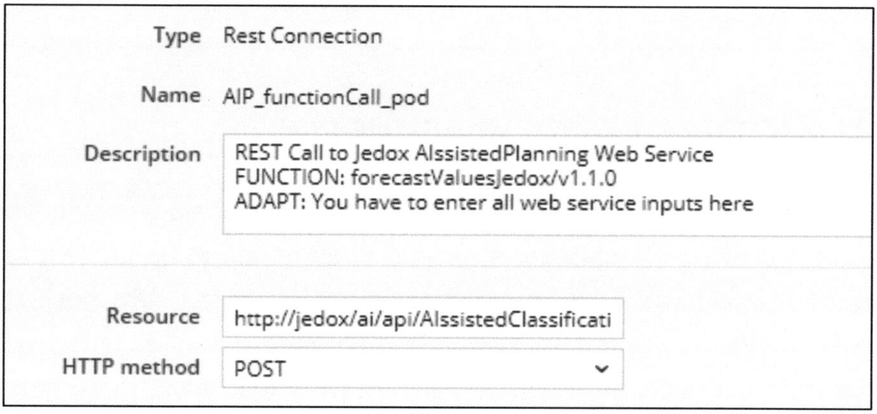

You of course do not need to fill in every description in every component, but you should provide descriptions for any unique or critical step.

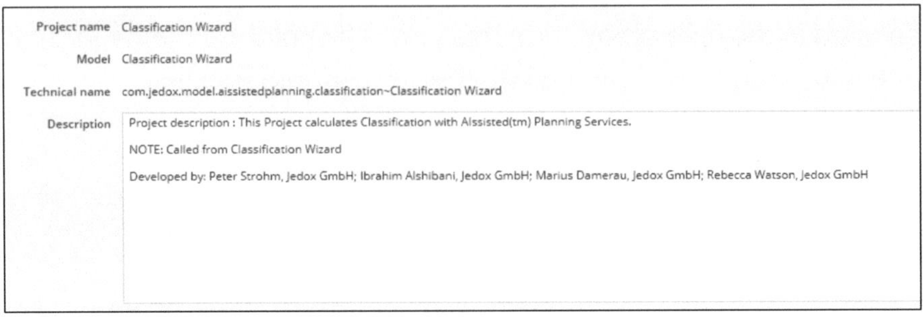

Once you have done this, the notes you have created will show up in the project documentation automatically.

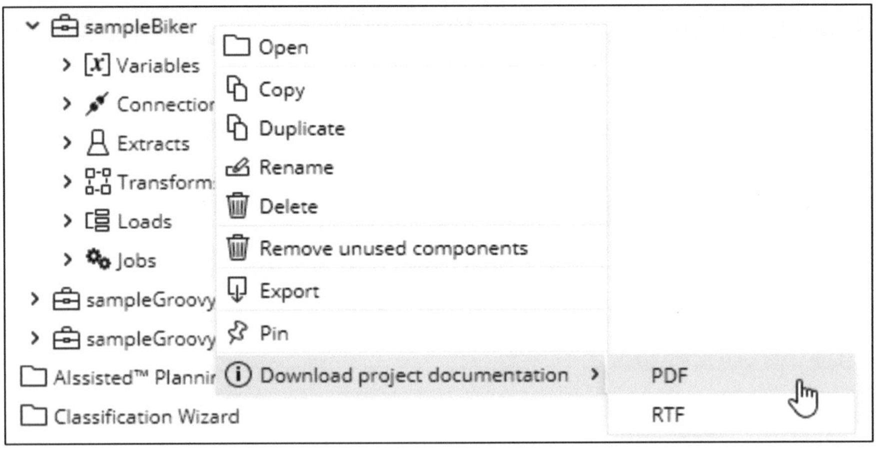

STRUCTURE LAYER 4: SECURITY

Like the general design of the model, security design should be brainstormed on a whiteboard or in a document and be subject to careful consideration before you implement it in Jedox. Why not just dive in with security? The reason is that being a multidimensional database, security in Jedox can get complex fast. Also, there may be specific security protocols/thresholds within your organization that you want to ensure the model meets or exceeds. Aim for the simplest model that will satisfy all corporate security policies.

Research how security is set up in your organization and across source systems, particularly the ERP/Accounting system, and how Jedox objects may have a relationship to the ERP security. For example, an organizational structure dimension (e.g. Cost Center) might reflect the ERP security configuration. Most importantly, ensure a subject matter expert from IT and/or finance is engaged in the final decision-making process for the Jedox security model.

User Level Security

This is the level that controls the user access to the system. When designing user level security, think about how Jedox can leverage of existing security infrastructure and how it aligns generally with the corporate security strategy for user authentication. External validation systems like SAML connected to Jedox have the advantage of ensuring only corporate authorized users have access. When a user account is cancelled, suspended or removed centrally, all access rights will be automatically removed from Jedox as well.

Groups and Roles

When structuring your user level security, consider the grouping structure of your users. As mentioned earlier, this structure is usually mirrored within the organization. This can typically take the form of an organizational structure, cost center structure or divisional breakdown. Jedox manages user security by using groups, just like how Windows AD security works. Users are assigned to groups, and groups contain the security assignments. For example, you might have a regional structure you want to mirror in your security groups:

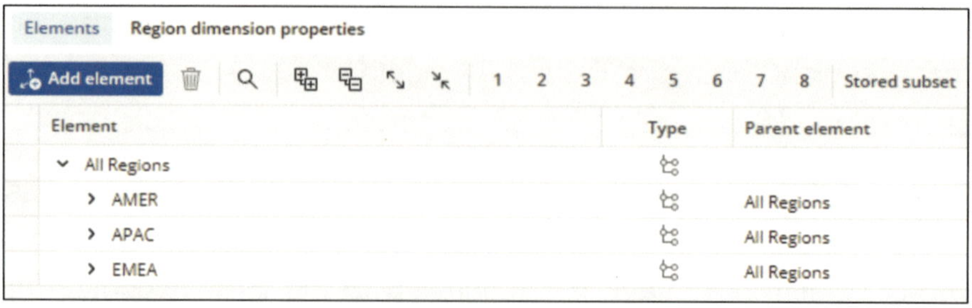

Your goal could be to restrict users to only see structures and data from each of these levels and below. To do that, you could set up the following 4 security groups in Jedox:

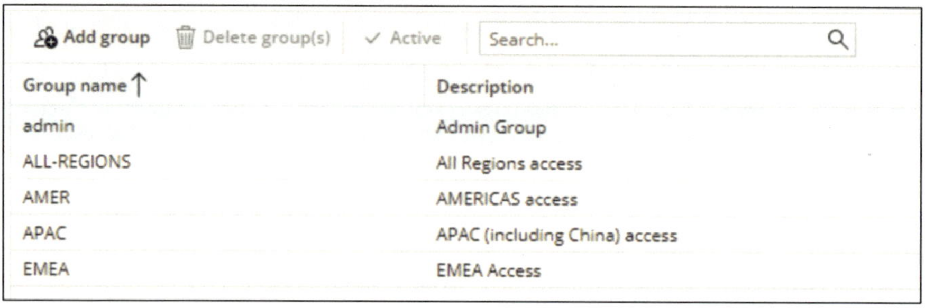

Because of the cascading nature of Jedox security, users with membership to ALL-REGIONS would have access to everything below (i.e., everything); membership to ASIA would only give access to ASIA related objects and everything below. Actual access has not been set up yet, just the structure of how it would look like. What happens though, if you have users that require **read** access to ASIA, and some users that require **write** access to ASIA? To achieve this, you can use Roles.

Roles are essentially the characteristic of the security and allows the assignment of access to certain objects in the Jedox platform. If you think of Groups as controlling what data you can access, Roles then define what functionality you can access. The advantage of this approach is that individual model-specific security can be assigned to groups, while the more generic platform related access is left to Roles. Roles can then be assigned to Groups.

For example, the standard "editor" role that comes with an instance of Jedox allows reports access only and all other components of the platform are hidden. In addition, the editor role assigns cell data access to the group. Note, if you are giving splashing access to your editors, ensure you increase the cell data access to "Splash". The "viewer" role has the same platform access as the "editor", except cell data access is restricted to read only. As existing roles can be modified and new roles can be created, consider starting your security model with existing roles, and modify them as required.

In our example above, to achieve groups that differentiate between read and write, you need to set up corresponding groups. Each region would then have a read "R" group and write "W" group:

Add group | Delete group(s) | ✓ Active | Search...

Group name ↑	Description
admin	Admin Group
ALL-REGIONS-R	All Regions read access
ALL-REGIONS-W	All Regions write access
AMER-R	AMERICAS read access
AMER-W	AMERICAS write access
APAC-R	APAC (including China) read access
APAC-W	APAC write access
EMEA-R	EMEA read access
EMEA-W	EMEA write access

Each group can then have the appropriate role assigned: “editor” for write groups:

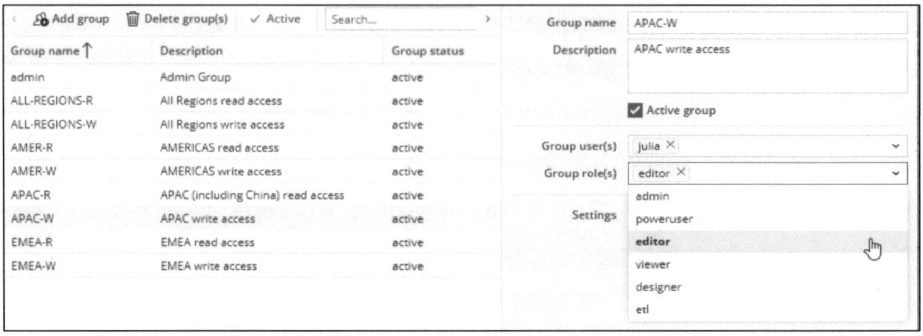

“viewer” for read-only groups:

This is what the corresponding editor role looks like with cell data access:

...and the viewer group with no access to cell data:

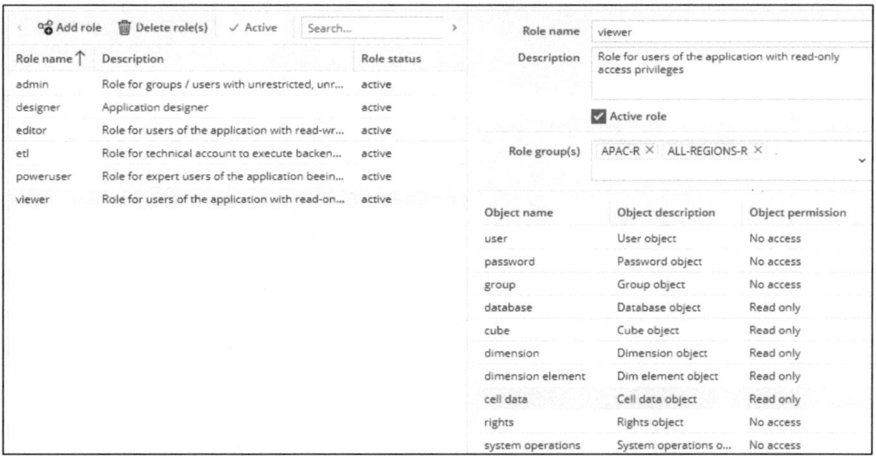

Groups and roles have description fields. Make sure you use these, especially if there are a lot of different groups to help with identification. The search function in the user administration area will also search across descriptions, so providing meaningful descriptions is super useful. This gives the possibility of creating a structure in the description field to provide further differentiation of users when searching.

Now groups and roles have been established with the users, we can look at assigning access to the specific, custom objects within our application.

Object Level Security

Once a user has been authenticated, the next step in the security process is to give them access to the various objects in Jedox from your model. This type of security is hard to outsource to a third-party system, as the objects (like reports, templates, cubes, dimensions) are unique to Jedox. Let's look at the various objects that you might want to consider for security.

Report Access

This allows you to restrict access to reports when they are used in the report tree in the reports section of the platform.

Note, report level security does not restrict access to the report if you are just linking to it via a URL - it merely restricts access to the node in the report tree:

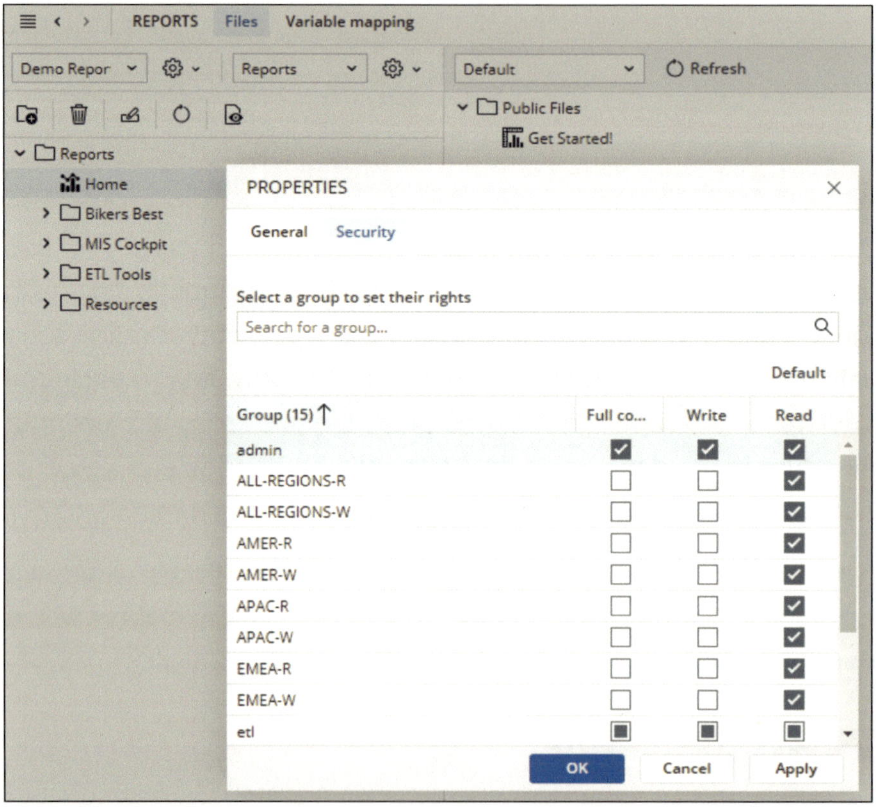

To access the report properties, ensure you turn the report user preview off:

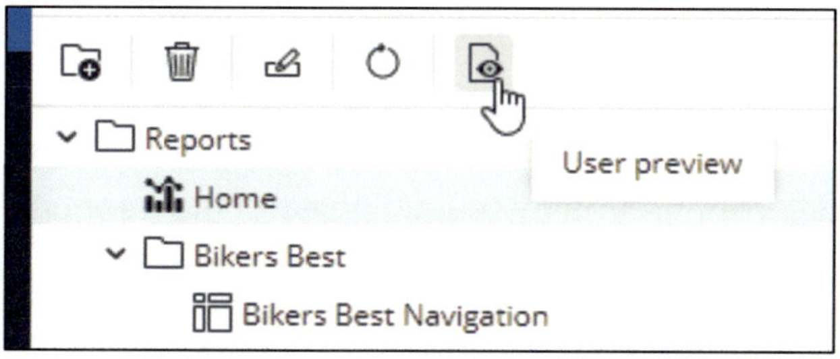

To completely restrict access to the report, you need to remove access in designer:

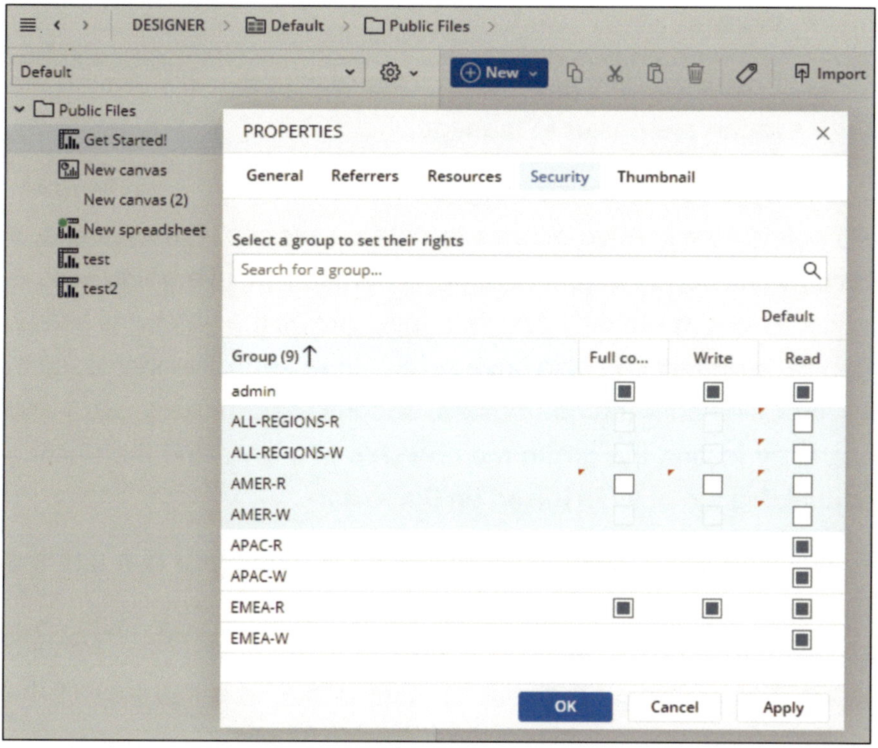

This will completely hide/restrict access to any method of accessing the report.

Restricting Objects on Reports

Often, the goal is not to hide the whole report. Instead, it is to use the same report for different groups by restricting what the user can see or do with it. This approach has the following advantages:

- Easier report maintenance. You need to maintain a single report, as opposed to many reports for different users.
- Consistent user experience. All users will see the same report, but just restricted to specific data and objects for them.
- Easier security application. Changes to security around the report are easier to manage.

If security is set up correctly, group membership for specific users will automatically drive access to data on reports. For example, if security has been assigned to a region dimension, users with access to just EMEA would only see that node and all the elements below. Report designers can also drive access to report components, such as form elements (combo boxers, buttons, date pickers, etc.) and charts by testing the group membership of a user and then hiding or unhiding the objects based on the result.

To test a user's access to an object on a report you can use the MATCH() and INFO("usergroups") spreadsheet functions. This allows you to check whether a user is a member of a particular group. For example, to check whether a user is a member of the "sales" group, you would use the following function:

=ISNUMBER(MATCH("ALL-REGIONS-W", INFO("usergroups"),0))

Using ISNUMBER() function is useful as it returns a Boolean (TRUE/FALSE) that you can then use in the visibility property of a chart object, for example. You can then use the result of this function to change the visibility of a form object on the worksheet:

Cube and Dimension Access

To ensure that access is applied correctly regardless of how the user is accessing the data

(e.g. via a report or paste view), ensure you define and restrict access within the actual database. In the cube properties in Modeler, you can restrict whether a group can have read, write or no access to a specific cube. All Jedox object security is held in security cubes and dimensions, so you can access the underlying security data directly.

When designing cube, cube data and dimension element level security, only enter security values where it is specifically required. If a security cell is left blank, it will inherit whatever security is assigned above it. For a detailed explanation of Jedox security, please see https://knowledgebase.jedox.com/jedox/security/rights-users.htm

In Jedox there is no similar security concept for dimensions. To have access to a cube, it implies that you have rights to the underlying dimensions assigned to that cube. You can, however, restrict dimension elements, as we discussed above with the region example.

Cell Data

This is the lowest level of security that can be achieved in Jedox - literally restricting access to individual cells on cubes. Although cell level security is not inheritable, you can define it with rules, which makes it a lot easier to maintain and use.

Typically, you want to use cell data security when defining security across dimensions. For example, you might want users to have write access to Budget in 2023 but read-only to Budget 2022:

Product Budget

Version dimension	Budget	Budget
Year dimension	2022	2023
Product A	READ	WRITE

STRUCTURE LAYER 5: CLOUD INFRASTRUCTURE

Given that the cloud console is built to deliver information about your environments, there is nothing to optimize from a user perspective.

If you have a lot of administrators of your Jedox cloud instance, it is important to have regular reviews of user access. This ensures only authorized users have access. Do not leave security holes by forgetting to remove ex-employees from the list of administrators. What can help is setting up multi-factor authentication. Once an employee leaves and loses access to their device, they will no longer be able to log in.

STRUCTURE DESIGN: IN SUMMARY

- Design away from the keyboard
- Be consistent with naming conventions
- Use a global workbook
- Keep security simple

Chapter Eight

Design Performance: The Model must be Fast

> ***"I feel the need... the need for speed!"***
> *(Top Gun)*

After 20 years in the business of enterprise reporting and planning technology, after data accuracy I have observed the top three design priorities for **users** always boil down to:

1. Performance
2. Performance
3. Performance

.. and **all other considerations** (in the user's eyes) are a distant second. Speed is by far and away the most sensitive design characteristic. But it is one of those that is only recognized by its absence. No one complains about a fast report.

There are a huge number of factors that determine a model's performance. From a user's perspective, performance normally means how long it takes for them to perform a specific action. For example, to refresh a report, log in, or enter numbers. **Fundamentally, the more the system must do to calculate, the larger the performance impact will be. Our goal is to build the model with the most efficient calculations possible.**

It is useful to set performance benchmarks in your project. This helps manage users' expectations and sets a clear standard that the model must meet before going live with the project.

Example Benchmarks (agreed with users)

System login to the first menu: **1-5 seconds**

Opening any report or planning template: **1-5 seconds**

Recalculating a report or template: **1-5 seconds**

Entering numbers, moving to the next cell: **1-2 seconds**

Max load time for refreshing daily numbers: **10-15 minutes**

Note, the above is an example only. Individual models and expectations will always vary. It is useful to have this as a guide while developing but then revisit and adjust the metrics (if necessary) on completion of the build phase.

Once you have determined the standard that users are prepared to accept and are reasonable, how do you build your Jedox model to meet these expectations?

PERFORMANCE LAYER 1: USER INTERACTION

This layer is where any performance issues are showcased - the good and the bad of your model is exposed to the users. Let's review again the key components of visualization in Jedox and how they are affected by performance.

Visualization concepts in Jedox

Jedox Reports. This component includes a spreadsheet-like designer to build interactive and dynamic reports and data entry templates. It is a cell-by-cell report building tool which is fully controlled by the report builder. This is the most popular method to build user content in Jedox. Reports allow nearly any type of structure and formula. Because you can combine any number of dimensions, cubes, databases onto a single report, they are at the most risk of being impacted by poor design choices.

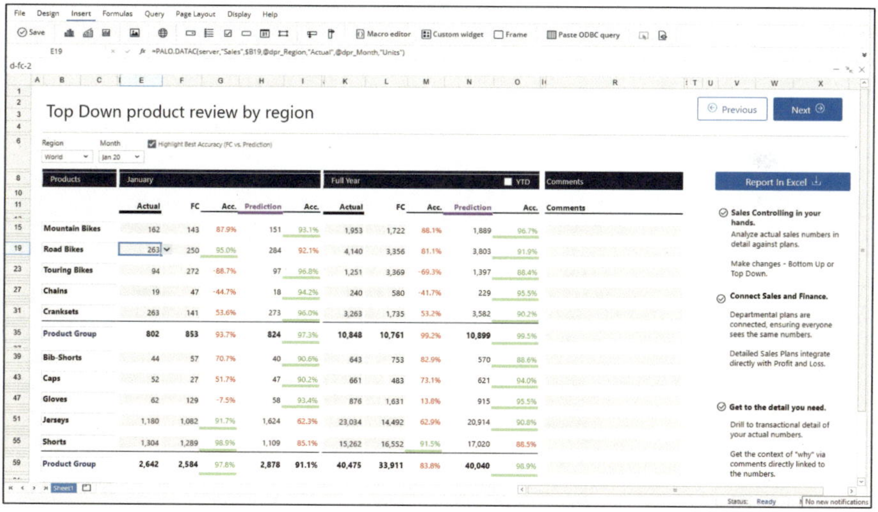

The advantage of Reports is that cell-by-cell content creation, and the report builder can control exactly how the user interacts with the report via dynamic objects on the report.

Reports can include (amongst other things):

- Dynaranges
- Conditional Formatting
- Macros
- Action buttons
- Writeback cells
- Custom formulas
- Conditional hidden column/rows

Views are direct database objects that can be displayed directly as a report. All data, structures and orientation of the view is managed by the database.

In addition, view themes allow structured formatting of the view, including cell locking and unlocking. View lists allow further flexibility to make views report-like.

Sales Report

World · 2019-03

	2019-03 Actual	2019-03 Budget	2019-03 Variance	2019-03 YTD	2019-02 Last Month	2018-03 Last Year
Bikes	241,660	157,336	-84,324	769,873	253,997	179,267
Handlebars	21,349	5,242	-16,106	39,950	5,953	9,631
HL Bottom Bracket	4,198	1,643	-2,556	10,349	3,075	948
LL Bottom Bracket	2,468	1,177	-1,291	7,978	2,468	573
ML Bottom Bracket	2,771	1,125	-1,646	7,678	2,226	632
Bottom Brackets	**9,438**	**3,945**	**-5,492**	**26,005**	**7,769**	**2,153**
Front Brakes	1,145	369	-777	2,920	995	714
Rear Brakes	1,199	415	-784	3,754	1,678	750
Brakes	**2,344**	**784**	**-1,560**	**6,674**	**2,673**	**1,464**
Chains	607	925	317	1,879	658	490
Cranksets	9,222	3,251	-5,971	29,041	10,038	1,762
Front Derailleur	25,010	18,278	-6,732	74,358	21,339	10,041
Rear Derailleur	13,604	11,406	-2,198	45,502	16,096	5,554
Derailleurs	**38,614**	**29,684**	**-8,930**	**119,860**	**37,435**	**15,594**
Forks	29,760	5,531	-24,229	54,365	10,066	11,545
Headsets	2,176	2,062	-114	9,222	3,742	1,659
Mountain Frames	67,606	19,481	-48,125	136,738	51,544	50,408
Pedals	4,465	8,507	4,043	12,273	4,001	3,565
Road Frames	42,215	16,141	-26,074	104,528	34,269	31,513
Saddles	4,218	7,929	3,711	12,057	3,674	3,362
Touring Frames	4,837	7,572	2,735	12,722	4,132	3,810
Wheels	14,653	10,003	-4,650	34,356	10,059	11,102
Components	**251,502**	**121,057**	**-130,445**	**599,670**	**186,013**	**148,057**
Clothing	129,499	183,297	53,798	365,842	150,504	102,954
Accessories	82,674	18,667	-64,007	151,851	20,565	133,038
All Products	**705,336**	**480,358**	**-224,978**	**1,887,237**	**611,080**	**563,316**

The biggest advantage of views is that they are server-based and fast. There is no processing happening at the user side, so they render within seconds. You can also access the same view in Excel and Web.

Canvas Is a dynamically scalable report object that allows data, visuals, and controls from various reports (including data entry cells) to be brought together and composed as a single dashboard-like visualization.

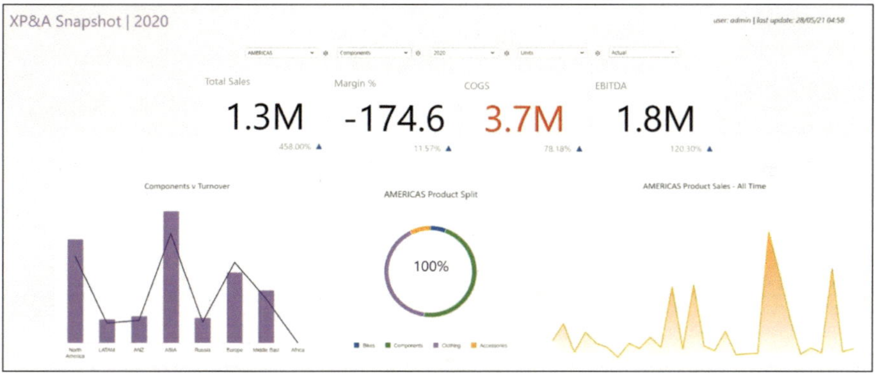

It is smooth scrolling and re-orientates based on the user device. Currently, only objects from existing reports can be used; you need to build your content first on a report and then drag it onto a canvas.

A canvas can be published just like a report. Both views (embedded in workbooks) and reports can be dragged onto a canvas. Report performance can be improved immediately by dragging it onto a canvas to take advantage of its smooth scrolling capabilities.

The Jedox Excel add-in allows users to build reports, templates and do analysis against Jedox data directly in their favored and familiar environment, Microsoft Excel. It gives the full features and power of Excel, while adding security, structure, and governance to the data.

Below is an example of view in Excel; you can of course build cell-by-cell reports in Excel as well. Excel is great for large reports (deep and wide), and when the users need to connect to existing spreadsheets and send data into Jedox.

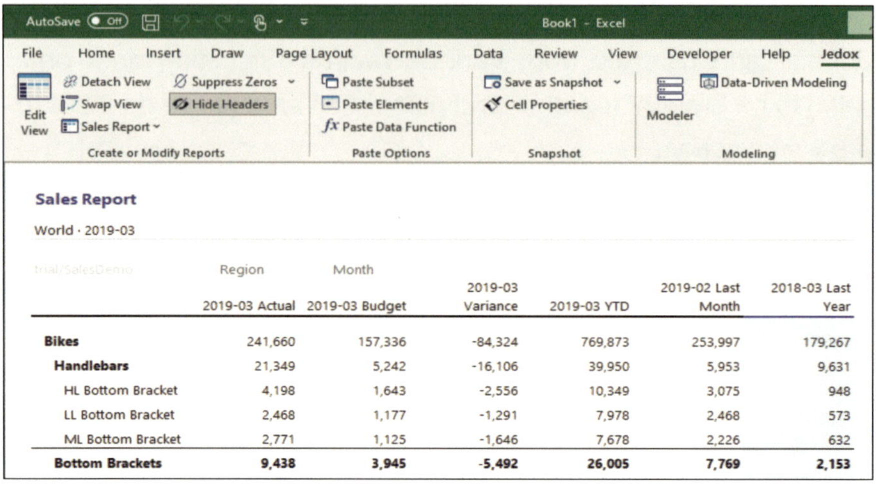

	2019-03 Actual	2019-03 Budget	2019-03 Variance	2019-03 YTD	2019-02 Last Month	2018-03 Last Year
Bikes	241,660	157,336	-84,324	769,873	253,997	179,267
Handlebars	21,349	5,242	-16,106	39,950	5,953	9,631
HL Bottom Bracket	4,198	1,643	-2,556	10,349	3,075	948
LL Bottom Bracket	2,468	1,177	-1,291	7,978	2,468	573
ML Bottom Bracket	2,771	1,125	-1,646	7,678	2,226	632
Bottom Brackets	**9,438**	**3,945**	**-5,492**	**26,005**	**7,769**	**2,153**

What Approach Should I Take?

Generally, Jedox web reports are the best for controlled, process-driven delivery of reports/templates to a wide user base for performance. You can control user workflow, template design and security much more easily than you can in Excel. Excel is typically for “power users”, or those users that require heavy analytics, or combining standard excel data with their Jedox reports.

Always balance functionality with performance. It’s tempting to cram a huge amount of functionality into a “one size fits all” report using the flexible Jedox spreadsheet capability, but you run the risk of performance bottlenecks and overwhelming the user with content. Instead, remember you can easily link reports, and chained analysis can be much more intuitive.

Consider using views for reporting and data entry as they are dynamic, and server driven. They are very fast to render, but the trade-off is that they have a limited array of customizations available from a reporting perspective. Each view is restricted to one cube only. You can customize the theme (the colors, font, and row height of sections of the view), but you are restricted to:

- One view per tab
- All layout aspects (rows, columns, data, titles) are controlled by the server
- Combo boxes can be used, but variables must be set up in the view first
- Cannot use actions or other form objects from the spreadsheet

Ensure you are only using what is necessary when designing for performance. Be efficient. When building your reports, keep testing against the defined benchmarks to ensure you are still within the performant zone.

Existing Reports. Where do I Start to Fix Performance?

The troubleshooting guide below is for a project that has delivered their reports, but still find the response time unsatisfactory. Some performance bottlenecks may be at the report level, but it also could be exhibiting symptoms of issues with deeper layers below. Here is a list of common visualization-related issues on the Jedox spreadsheet and potential ways to solve them:

ISSUE	POSSIBLE CAUSE	POSSIBLE SOLUTION
Opening of a report is slow	Physical size of the report is too big (check file size in Designer).	Open report, check to ensure there are no blank rows underneath the last row on the report (to check, select last row and click CTRL end on your keyboard. If there are any blank rows, delete and re-save the report. Check the file size again. It should now be smaller.
	Too many large images in report	Ensure all images have been edited and resized appropriately before import. Just resizing the image on the spreadsheet will not reduce the image byte size.
	Initial state of Dynarange is too large on report open	Look to change the initial open state of the Dynarange to one that returns a small quantity of rows initially. If this is not possible, look to change the Dynarange to a subset that utilizes paging. This will enable you to set the initial opening to a controllable number of rows.
	Reduce or simplify the Dynaranges on the report	If you have many Dynaranges on the report, try replacing them with static subsets of elements or values driven by normal spreadsheet formulas. If the report is from one cube, consider replacing with a view.
	Reduce the number of tabs in the report	Jedox will attempt to calculate all related tabs to ensure dependencies on calculations are refreshed. This can slow down the initial opening of a report. Try to keep the tabs to a minimum and use linked spreadsheets (tagged as a resource) where possible.
The combo boxes take a long time to appear	The dimension subset or other data driving combo box is too large	Edit each combo box and ensure that the "Large subset" checkbox is checked. Review the subset or SQL Query driving the combo box and see if it can be optimized.
The report is opening fine, but charts and tables of data are taking a long time to appear	Complexity of calculation to resolve objects on the spreadsheet	This can occur when there are heavy calculations on the spreadsheet before the final data is rendered. For charts, try to use views directly as a data source; for other data, the to simplify the calculations and Dynaranges that are driving the displayed numbers

Note, the above guide is not a complete list! Reports can be very complex and there can be many reasons for performance impacts. You can see more information on the Knowledgebase under "Best Practices for Jedox Spreadsheets" https://knowledgebase.jedox.com/jedox/jedox-web/jedox-spreadsheets-best-practices.htm.

In release 2022.4, Jedox introduced **Spreadsheet Advisor**, which can help identify bottleneck components in complex workbooks.

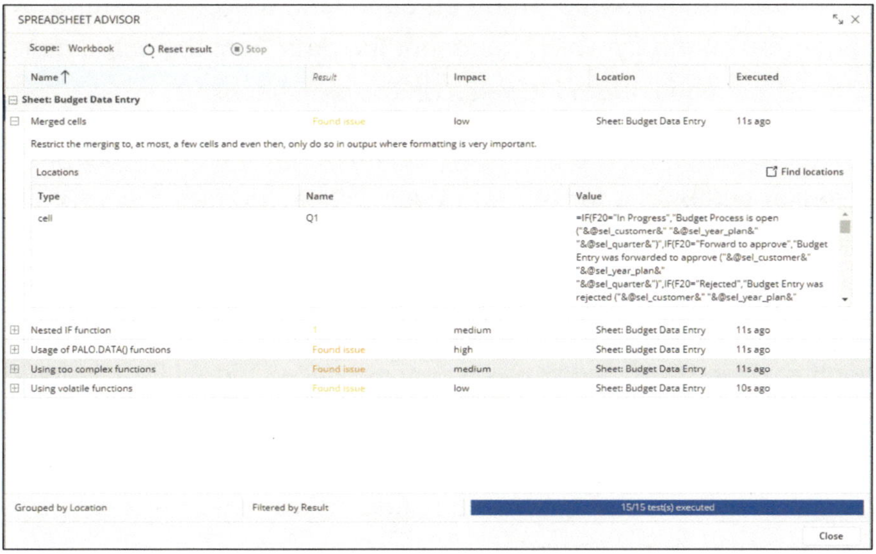

The Spreadsheet Advisor allows you to select various tests and run them against the spreadsheet. Each test is weighted with the performance impact, which allows you to sort and select those which will have the biggest impact on your spreadsheet.

PERFORMANCE LAYER 2: DATABASES (CUBES, DIMENSIONS, LISTS)

Layer 2 is essentially the "engine room" of your model. This is where the structure, data, calculations, and shape of your model lives. Getting this wrong can have major knock-on effects for the performance by-end users. The general rule is **the less calculations for the server, the faster the server will respond to requests**.

Cubes

Performance in a cube is not so much dictated by the order of dimensions, more so by the quantity of them. This is because to return data from a cube to the spreadsheet, formulas need to be generated on the fly which represent the data from the cube; more dimensions equal more formulas on the worksheet.

The more dimensions a cube has, the more options (therefore more potential confusion) for the end user. While you can theoretically add 100 or 200 dimensions to your cube, practically, the user will get lost. Consider each dimension carefully in turn, and whether it could instead be included as an attribute or a secondary hierarchy within another dimension. Jedox has a concept called Virtual Dimensions, which allows you to turn any attribute into a dimension dynamically by the server. This will also reduce the number of potential cells in a cube and a smaller cube means less server calculations, which means better performance.

The biggest potential performance impact lies on the server with rules. Rules are in-memory calculations on a cube that occur in real time (i.e., as the user requests them) and are analogous to excel cell formulas. They are generally easy to create and are extremely powerful: they allow you to do on-the-fly calculations against entered cell data. However, having inefficient rules where the calculation scope is across a large part of the cube can cause the database to slow down and in extreme cases take a long time (minutes to hours) to resolve calculations.

We will not deep-dive into rules here, but here are some performance related key points you should think about when designing rule calculations for your model:

- **Is the data volatile?** A general rule for good performance: use rules for volatile (i.e., user entered or fast changing) data, and use Jedox Integrator to calculate the rest.
- **What dependencies am I creating?** For example, if you have a very large sales model where your rules are not optimized and you then have another rule that pulls the results into a separate Profit and Loss cube, you may see a cascading performance impact.
- **Will the calculation be widely used?** If the calculation is used for a single user for example, could that rule become a spreadsheet calculation?
- **Can the scope of the rule be restricted?** You can be very prescriptive with Jedox about exactly where in the cube the rule will apply. Ensure you are only calculating cells in the cube that you really need. Generally, the more specific you are on the left had side of a rule, the more restricted the scope will be. And the inverse applies. For example, an innocuous rule that starts with []=If()... will actually be testing every single cell in the cube.
- **Am I using the fastest rule engine?** Are the rule functions I am using compliant with the fastest rule engine in Jedox? Ensure you are using rule syntax that is compatible with the Data Driven Engine. Check out the Jedox Knowledgebase "Rule Engines in Jedox" for more information on this.
- **Am I using element weighting?** Consider using element weighting on a consolidated element instead of rules to calculate simple (+ or -) arithmetic functions. The server is generally better at calculating consolidations than rule calculations. (Note: this is not always appropriate nor the best design; if many elements are created just for this purpose, a

simple rule might be more aesthetically appealing and less confusing).

Dimensions

Dimensions themselves can have a large impact on cube performance. The essential point is to build only what you need. Here are a couple of tips to ensure optimal performance with your dimensions:

- **Do You Really Need all those Elements?** I have seen dimensions examples containing millions of elements when only a few thousand or so are used. Think about exactly what you need as part of the integration process. Unused elements can create a completely unnecessary calculation overhead, especially when trying to view the elements in the dimension or creating a view.
- **Restrict the Attributes on the Dimensions to what is Required.** Attributes are stored in a separate system cube and are calculated when the dimension is calculated. Ensure that you only have exactly what you need for your model.
- **Dimension Depth.** Manage your structures so that you are efficient with the depth. The deeper a hierarchy is, the more calculations may have to take place on the server.
- **Multiple Hierarchies.** This is a powerful feature of Jedox, but too many can cause confusion and slow the initial calculation of a view down. Again, use only what you need for your project.
- **Set Default Elements.** Set the default parent, write element, read element, total element and na element in the dimension properties, where appropriate. Setting the default read element for example, will mean that element

will be displayed by default to the end users (assuming no security setting overrides this). It can mean that the element selector in the spreadsheet can be more efficient, faster and more user friendly.

Lists

Lists are a server-side object that is defined by the user. Essentially, you can define intersections in a cube that exists across dimensions in a way that is impossible using dimensions alone to do.

Technically, stacking dimensions gives you a cartesian join which returns all possible combinations of the elements in those dimensions. Lists give a non-cartesian join as the user has defined exactly what should be returned. Lists currently only work with views, and you can only have one list per view. When looking to improve performance, use views with lists to replace reports where possible.

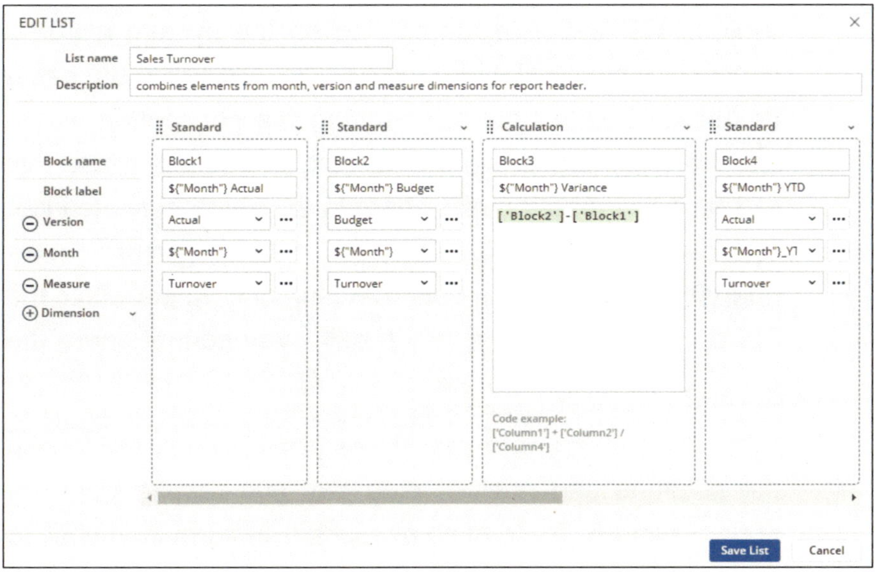

PERFORMANCE LAYER 3: INTEGRATION

Most processes built in Jedox Integrator do not directly affect end-user experience, but they can still be bottlenecks to overall system performance. It is common for a model design to sometimes have users executing ETL processes from reports. In that case, efficiency is critical. Out of the three steps in the Extract-Transform-Load process, typically most performance issues occur at the Connection or Extract stages. This is where Jedox Integrator is requesting data, typically from a third-party system. I will not cover every single extraction type in detail here, but here are a few key things to look for if your integration performance is slow.

- **Connections.** Ensure the connect and authorize method you are using is the most efficient one available for the source system. Sometimes, there are multiple approaches that can be used with the source system for connectivity. Ensure the one you are using is the most efficient for that system. For example, depending on the system a JDBC connection may be faster than a REST API call to the same database. Ensure you are caching the connection where possible. By ticking the cache box, Jedox will reuse this extract if it is required elsewhere in the same executed job. This can save tons of time as it prevents Integrator from re-querying the underlying data. Caching only works within the frame of a single top-level job call - you cannot share the cache to another separate job.
- **Extracts and Transforms.** Be as precise as possible with the columns and rows of data that you require from the source system. Try to push as much transformation back to the relational database as possible as SQL query statements in the extract. This will make your downstream Jedox

transformations simpler and will have less overhead when loading the data later. If you tick the cache icon, remember the data is cached once and reused per execution. If you want to force Jedox to recalculate the data, simply do not tick the box.

- **Jobs.** Take advantage of parallel jobs where possible. This can have a huge impact on loading time. I have seen customers reduce a 7-hour load job down to 40 minutes, just by smartly parallelizing as many jobs and loads as possible. Depending on your integrator projects, this can be a complex and time-consuming operation. Knowing what to run in parallel and what to run in serial may be critical to the accuracy of your data. The great thing with this point is that it is completely in your control. Regardless of the performance of the source system, you can improve it by processing loads and jobs together.
- **Scheduling Timing for Jobs.** Sometimes the third-party system that the integrator is connected to has a scheduled dependency, meaning that there may be overnight processing that occurs before the data is available for Jedox. Sometimes, the timing of this can change daily and may conflict with the scheduled Jedox jobs. You can add a Jedox job (via batch files or OData execution URL) as a final step in the third-party processing job, so Jedox loads as efficiently as possible from a timing perspective.

PERFORMANCE LAYER 4: SECURITY

Jedox security can be divided into user and object level security. User security is managed via a centralized Active Directory another authorization system, or locally within Jedox itself. If users connecting to Jedox are finding the external authorization is taking a long time to occur, this is an exclusive task for IT security to optimize.

If using SAML for authentication, there is scope for optimization. If you have many users, you can use the SAML in Authorization mode. Only group-role assignments need to be defined in Jedox, and the SAML verification process handles the rest of the authentication. See the knowledge base for more information – SAML Authentication in Jedox: https://knowledgebase.jedox.com/jedox/security/saml.htm

Jedox also has a concept of a “superuser”: the admin user account. When using this account, after the login process no other checks are done and therefore it is the fastest authentication possible. This is common setup shared with many other types of databases. For this reason, it is crucial that any performance tests are made with “normal” accounts, as using the admin account may mask potential performance issues.

Object security is controlled within Jedox and can be a lot more complex and can go to cell level. As a principle, you want to design the lightest security “touch” possible, while at the same time providing comprehensive security for your model. Security (except for cell level security) is inherited in Jedox. That means, that if a parent element grants a user group read access, then members of that group will have access to that element, plus all related descendant elements.

Although security is optimized by the database engine, generally the more security restrictions you apply to your model, the more processing it must do to validate access, right down to an individual cell in a cube. By thinking carefully about your design, you can optimize the levels of security your model has.

PERFORMANCE LAYER 5: CLOUD INFRASTRUCTURE

The cloud infrastructure layer, represented by the Cloud Admin console is the deepest layer here, but potentially one where, when all else has been considered, will help in performance. Jedox runs in a Kubernetes cluster, and internally it manages all resource allocation (memory, CPU).

Jedox resource usage is not constant. Memory and CPU usage tend to trend to zero or close to zero, then increase quickly once utilized. It means that you need to ensure the resources are sufficient to cover the maximum levels of usage, not the average. In this way, Jedox infrastructure utilization is not unlike electricity infrastructure, with spikes in demand. Generally, as a rule of thumb, you require 2.5 times more RAM currently utilized as a buffer.

By contacting Cloud Support, you can get them to analyze your instance and see if the physical resources require boosting. They will be able to see clearly if the system is over-utilizing available resources and will recommend additional resources if necessary. This might mean additional CPU time or memory allocation for your environment.

This is generally only helpful if you are encountering performance issues across your entire system.

If your environment is scaling across multiple users (many thousands) or huge datasets and calculations (100's of gigabytes in memory), Cloud Support should be notified up front before your project starts, to anticipate the resource usage.

DESIGN PERFORMANCE: IN SUMMARY

- Use views for reporting and data entry where possible
- Calculate only volatile data with rules
- Perform processing in the relational extracts to avoid complexity
- Keep security design simple

Chapter Nine

Designing Scalability: Unleashing the Beast

> *"Scaling is multidimensional. Different methods address different things."*
> *(Fred Ehrsam)*

The Jedox architecture is designed for fast querying, planning and analysis. By default, the Jedox database is designed to hold a huge volume of data. It has a general compression rate of 10:1, when compared with standard relational databases. That is, a 10 GB relational table would generally equal 1 GB cube in Jedox. With most models the standard Jedox configuration and setup will suffice at running for optimal scale. Jedox is a database, and like any database, design is an essential component, especially when you are looking to scale the system up or out. Sure, you can always add more CPUs or memory to the instance, but the smart money is on how we use the current resources as efficiently as possible to scale.

What does scale mean? There are generally 2 dimensions of scale that can be considered:

- Scaling Vertically (Scaling up)
- Scaling Horizontally (Scaling out)

In a Jedox context, **scaling vertically** refers to how we can keep consistent performance and reliability of the system while adding more data to it. **Scaling horizontally** refers to how to keep consistent performance and reliability when adding more users. Scalability is closely related to performance, but it is not a complete overlap. Performance design is like a prerequisite or a consideration when looking at designing for scale.

When do I need to consider scale design in Jedox?

Designing for scale is not always a priority. Jedox naturally handles huge volumes of data. If you are building a small model and deploying to a small number of users, then spending a lot of time on scale design is not always required. However, if there is a possibility that in the future there could be a massive increase (i.e., 100 or 1000 times) in data or in users, then you should consider scale now. It is much easier considering scale at the start of a project rather than trying to scale later.

SCALING VERTICALLY (SCALING UP)

Scaling up starts with data. When data increases over time, the processing time to load that data naturally increases. Load only what is necessary and optimize for performance.

1. **Load only the data that is new or that has changed.** For complex models, this can be difficult, and will require thought and planning to achieve. For example, you might

have an Inventory Stock cube that contains opening and closing balances. The source system can re-state the data up to 10 days prior. This means that you need a process to recalculate stock values, including opening and closing balances for a rolling 10 days. It might be easier to reload all history every night, but the processing time is long. By only loading those 10 days of changing data, you will cut down on the nightly processing time.

2. **Segment historical data**. For huge volumes of data, keeping it all in the same cube risks slowing down the nightly process. If you can identify data that is not changing, segment that data into "archive" databases. For example, you may want to keep the current 2-3 years "live" in the current database and segment the previous 10 years to an "archive" database. The downside is that users will have to create different queries to create reports across both databases. But if the data structure is identical, it should not be that difficult to do.

3. **Segment current data**. Sometimes the volumes are so huge that it might make sense to segment current data into separate databases. This helps "quarantine" processing by the server and ensures user queries are spread out over the server. Again, this option is suboptimal when users need consolidated data, as multiple databases will be required in the reporting.

4. **Rewrite queries in Integrator extracts.** As the data load profile changes, original queries might not be appropriate anymore. Consider rewriting the queries, changing to a faster driver, or querying only what is required in the most efficient form possible.

5. **Review and optimize the transform steps in your project.** A simpler ETL project means less steps to move through which means faster processing times.

6. **Restructure source system data.** Depending on the system you are pulling the data from, it might be possible to optimize objects in the source database. For example, if loading from a data warehouse, is it possible for the source tables/views to be indexed/optimized, or temporary tables created to improve Jedox load times?

7. **Review model calculations**. For extreme datasets, replacing as many rules as possible with "hard coded" values will ensure the rules engine is only calculating what is necessary. This also means less calculations for the server to do. For example, archived datasets could have all their rule values replaced with ETL calculations, or replace some calculations on the spreadsheet with spreadsheet level calculations.

8. **Infrastructure**. If the data volumes are increasing, increasing raw power to your environment can help keep the user experience consistent. Contact Jedox Cloud Support for a review of your system and they will make a judgment if more resources (CPU, Memory) will assist in helping you scale.

9. **Integration Architecture**. If you have a time constraint on loading data into your model on top of the normal volume constraints above, you could consider a multi-integration environment. This would be a set up where multiple integration instances load different data into your model simultaneously.

Anecdote time: there was once a Jedox customer who had a time constraint of parsing over 900 million rows of relational data, performing calculations and loading the results in a 15-minute window. They achieved the load window by having multiple integrator instances loading data simultaneously. NOTE: This is an extreme setup - understanding your data load is critical in this environment to ensure there is no overlaps or other accidental deletion of data. This integration structure is nonstandard and would have to be set up by Cloud Support.

SCALING HORIZONTALLY (SCALING OUT)

A large increase in users has big implications on your model. Even if your model and data is relatively small, a huge user base can have an impact on performance. By huge, I am talking about thousands of users at a time accessing the system.

1. **Review the user reports.** How efficient are the reports? Are they optimized for minimal load time and recalculation times as much as possible? Remember, the less calculations reports need to do, the faster the reports will go. Can you change some reports to Views with Lists? Are users running integration processes from the reports? Is this causing bottlenecks? Is the report performant enough for the single "admin" user? Is it performant enough for a normal user? Remember, when testing report speed for a single user, the performance you experience will be the absolute fastest the end users will see. It will only get slower with more users and more data.

2. **Check your user-driven integration processes**. These are the integrator jobs and loads run by users on reports. Make these integration processes as efficient as possible, and always run them in parallel if you can. This will ensure multiple users do not lock up the system by queuing a lot of loads or jobs.

3. **Check your rules**. Every time a user touches a cell on a report that is connected to a rule, if that exact intersection or area is not cached Jedox will try to recalculate it. If there are thousands of users touching not one but hundreds (or thousands) of cells, depending on the calculations required Jedox could end up calculating continuously. Think about moving user calculations to the spreadsheet, so only the spreadsheet is calculated, not the server. Additionally, check if these calculations can benefit from Integrator. Some rules are difficult to transfer to an integrator job, but the calculation benefits might be worth it. Of course, some rule-based calculations cannot be replicated in Integrator (for example, consolidated level calculations), but it might be an opportunity to use cell-based calculations on the reports.

4. **Geographical considerations.** If you have a global planning model, optimization of the scaling out process can be a complex topic. Obviously, the simplest way of delivering is a centralized model that all users log into. However, local latency may mean that for some users, the experience might be suboptimal. The challenge with a planning (read/write) system is that you want to centralize the data, so changes are reflected immediately at the top level of the system. If a model is purely reporting only, a star/planet style database where there is a local server reflecting the main server, is relatively easy to deploy. This is where the "star" system

(i.e., the main/centralized system) does all the integration and nightly updates.

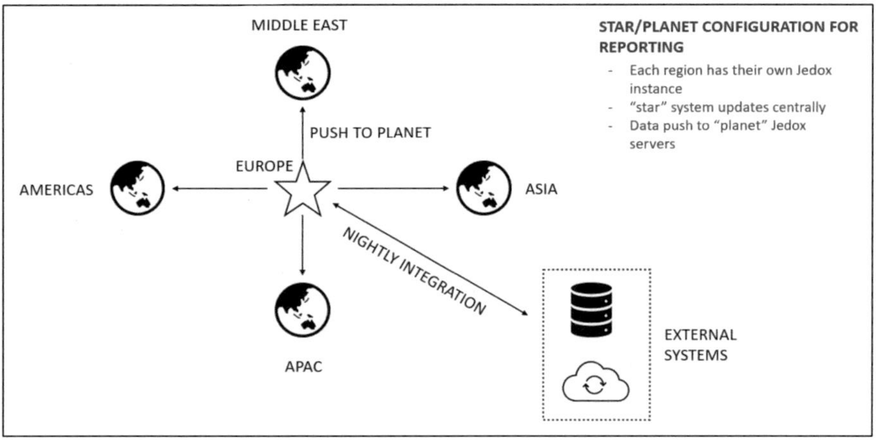

This is possible to do with Integrator and ensures all data is in sync.

With planning data however, options are more complex. The Planet databases could sync the data against the "Star" server:

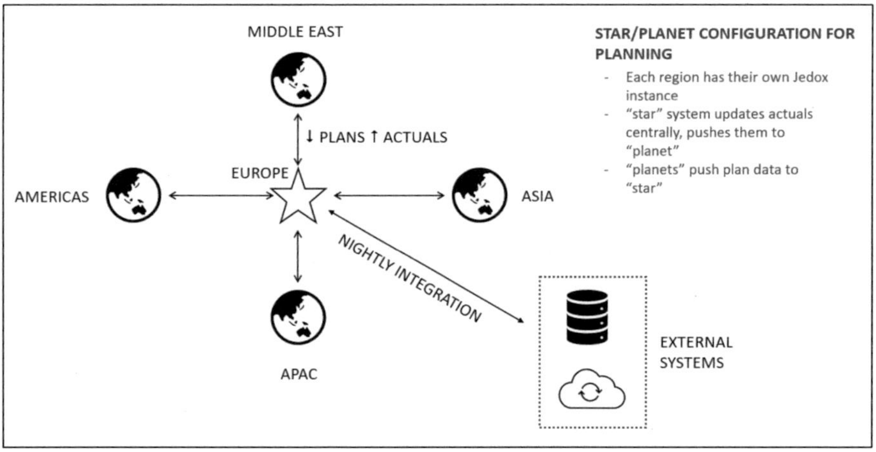

This configuration ensures that local users have the best performance possible. This only works when planning data is quarantined between regions and local reports do not have to see top level consolidations, or planning numbers from other regions. The data is consolidated at the group level via an integration process – regional users work on their own specific "slice" of the data. Note, the data does not have to be strictly segmented by any region or geographical dimension – it can be split in any way that makes sense for the model.

This also restricts top-down processes, like splashing. Bottlenecks can move from one part of the platform to the other when trying to solve complex latency issues.

A far simpler solution to geographical scaling is using an infrastructure solution. If your model has already been built without scaling considerations, and you are finding that distance is causing major latency for users, perhaps using Remote Desktop Services for the user front end can be the answer.

This will allow a consistent user experience regardless of their location. This will need to be provisioned by your IT department and will require their involvement to do this.

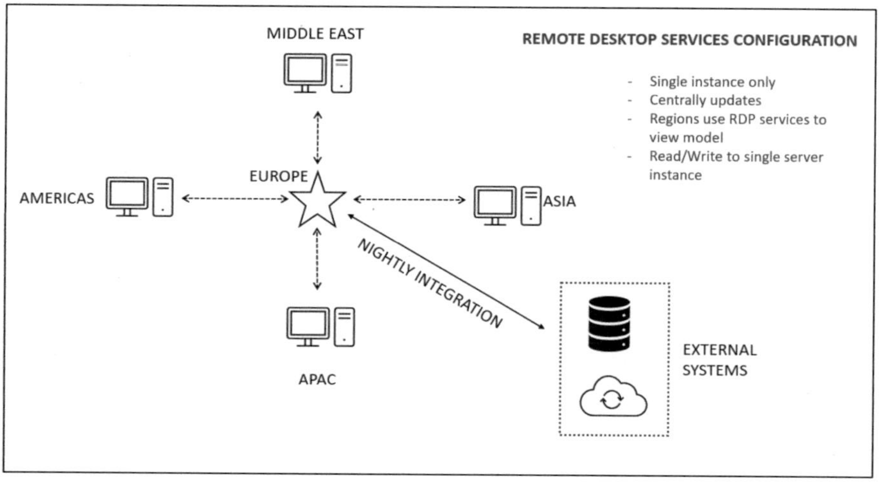

SCALING: IN SUMMARY

- Only load data that you need
- Segment your data where it makes sense to do so
- Ensure you cloud resources are appropriate
- Make all reports as efficient as possible

Chapter Ten

Designing Aesthetics: Making it Look and Feel Easy

> ***"Design must be functional, and functionality must be translated into visual aesthetics without any reliance on gimmicks that have to be explained."*** *(Ferdinand Porsche)*

How your model looks has a direct relationship to how it will be perceived by your users. Regardless of how much thought, hard work and feedback you have had on your model design technically, if it does not look attractive no one will want to use it. It's that simple. Humans are visual animals, and how something looks has a major impact on how we see things. Consider the look and feel upfront.

The double-edged sword with Jedox is that it is on one hand incredibly flexible. On the other hand, this flexibility means that some things like visual design need to be built from the ground up by you, the developer. Recent releases of the software have addressed this with canvas and view themes, but it can still take practice to generate beautiful cell by cell reports out of the box, every time.

WHERE DO I START?

Aesthetics can be tough. Knowing what looks good and what does not is an innate skill some people just have, but it certainly can be learned. Even if you consider yourself an aesthete, if your organization has an internal UX/UI or web graphics or marketing team, I strongly recommend that you start here. These people design good looking things for a living and generally know what they are talking about. They are also usually responsible for how corporate systems and web pages look and feel. Involving them from the start of your project is important. This ensures that they are aware of your project and can allocate resources to you as required. They should also be able to provide you with corporate logos, images, colors , fonts and font sizes. Their main contribution, however, will be their valuable ideas and thoughts on how users should interact with the system from a best practice perspective.

If you do not have someone in your organization responsible for this, then you could consider hiring a web design consultant for the design phase of the project. There are also a lot of free online resources that can help educate you on what looks good. **The Interaction Design Foundation** https://www.interaction-design.org is an excellent resource for articles, courses, and assistance with aesthetics.

Jedox also has a separate knowledgebase website dedicated to design topics in Jedox: https://design-system.cloud.jedox.com/introduction/design-system. This is a fantastic resource on a granular technical level. It is always good to understand the corporate approach of your organization first, then use this as a "how to" resource.

GATHERING EXISTING COLLATERAL

If you have access to corporate logos, images, colors, you can put them into a central location. You could make a folder in Jedox Designer and store design collateral there.

COLORS

If you have a corporate color palette, select 2-4 colors and use these as the basis for your templates. The colors should be complimentary as well as having the ability to contrast. You should be able to get the RGB of each, so you can be precise with the color on your reports. Think about not only the block background colors, but cell borders, fonts, active/inactive cells, and input cells vs non-input cells.

FONTS AND SIZE

Try to use a single font family across the model. If you have a custom corporate font, technically you can import it into Jedox, but this is discouraged because some parts of the platform (e.g. PDF generator) will not use the font in certain situations, leading to inconsistency. It is better to select one of the pre-installed fonts that are guaranteed to work everywhere. Think about the style and size of the font you want to use and where you want to use it. Have a clear language of font sizes. For example, all titles are font Calibri size 16, subtitles on reports are Calibri size 14. Row and column headers and cells are Calibri size 11.

LAYOUT

Consistency is the key. If you can avoid it, do not constantly change what the user expects as row or column headers and selectors at the top of the report. The best layout is one that "knows" the user (via security) and provides all options without them having to make selections. The less the user needs to do, the better. An easy way to achieve this is to provide a well-designed, consistent menu either at the top or the side of the report for navigation. This is best achieved via Framesets or with Canvas.

READABILITY

The ease with which a user can read your reports is crucial. Readability brings together color, fonts, words, and layout and is essentially the composition of your report. Ensure your font size is consistent and appropriate for various screen resolutions. Color is a big part of readability. Try to keep the backgrounds as plain as possible (i.e. a type of white or the default white), and use the colors to accent the background, not overwhelm it. Eye strain trying to read reports will be another big turnoff for users. Finally, do not overwhelm your users with lots of words and instructions on your reports. Take advantage of tool tips for mouse hovers and if you really need, have a separate how-to document that the user can read separately.

NAVIGATION

Whether you use the left-hand tree control to manage the report experience, a standalone menu, or an embedded menu/report in another system, think about the easiest approach for your users to find the information they need. The question is: how can I get the user to what they need with the minimum of clicks? The best click is no click.

DATA MODEL

The aesthetics of the data model is an important consideration. Prepopulate as much metadata as possible. For example, when a user logs in, the system should know who they are, the cost centers they are responsible for, the period for which they are planning for and the templates that they require. All of this is possible using variables and actions on the opening of the reports or main menu.

By pre-populating a model with the budget data, or a prior forecast, users only need to consider what has gone up or down or is new instead of thinking about everything from scratch. For cell-by-cell reports, you can take advantage of data validation drop lists to give users options to increase by percentages, whole numbers, or other splashing commands without having to use those commands manually.

A prepopulated process can dramatically improve not only the overall model impression on the users (after all, budgeting and forecasting is typically not the most favorite part of their job…) but will speed up the time to complete the plan which means more time for analysis and insight. Pre-population is not always possible or even desirable (sometimes you want your users to start from scratch), but it is something that should be considered in making the overall process easier for users.

AESTHETICS: IN SUMMARY

- Involve internal corporate UX/UI resources
- Read up on what makes a good aesthetic
- Keep the colors, font, sizes simple and consistent
- Prepopulate the model for users where possible

Chapter Eleven

Panic Notes

> ***"Don't Panic."***
> *(Douglas Adams, The Hitchhiker's Guide to the Galaxy)*

Ok, you have landed here because you have either:

1. Naturally reached the end of the book and are curious about this section, or
2. You have some serious problems and need to get up to speed very quickly with Jedox.

If you answered 2, most likely you are new to administering Jedox, and the responsibility has landed suddenly. Let's look at responses to critical and non-critical issues:

CRITICAL ISSUES

If you have one of the following "system down" situations:

- The data is not updating
- Calculations are not working
- Reports not generating
- Users cannot log in

AND you are in the middle of a planning or reporting cycle

AND time is of the essence

Do the following:

1. Check your Jedox Cloud Console. Log in and ensure all components of the system are up and running.
2. Check the general status of Jedox Cloud at https://status.jedox.cloud. All traffic lights should be green for your region.
3. If you have done the above and still not resolved your issue, **Contact the closest certified Jedox Partner to you to perform a triage of the situation**. The partner will know immediately what issues are related to your model and which ones are related to the actual platform itself. Anything platform (i.e., software) related, Jedox Support can be contacted immediately via the Support Portal and the team there will escalate to the development team asap. Do not escalate model related issues to Jedox Support (unless it is a Jedox-owned Model from the Marketplace). This is like expecting Microsoft Support to fix a formula in an Excel spreadsheet. The Jedox partner teams specialize in understanding models and are your best for a quick fix, or at least a clear understanding on what is going on.

Quick Links:

- Jedox Cloud Console: https://console.cloud.jedox.com
- Jedox Cloud Status: https://status.jedox.cloud
- Jedox Support Portal: https://my.jedox.com
- List of Certified Jedox Partners: https://www.jedox.com/en/partner/

NON-CRITICAL ISSUES

Non-critical issues are basically all those issues you might encounter that are not classified as complete "showstoppers" but still require urgent attention. Let's look at the same list of issues again, but with the eyes of someone who had time on their hands to investigate the issues themselves.

THE DATA IS NOT UPDATING

In most implementations, Jedox Integrator is loading data from source systems into Jedox. If data has stopped updating, this would be the best place to start looking for issues. Go to Jedox Integrator and click on the Monitor button:

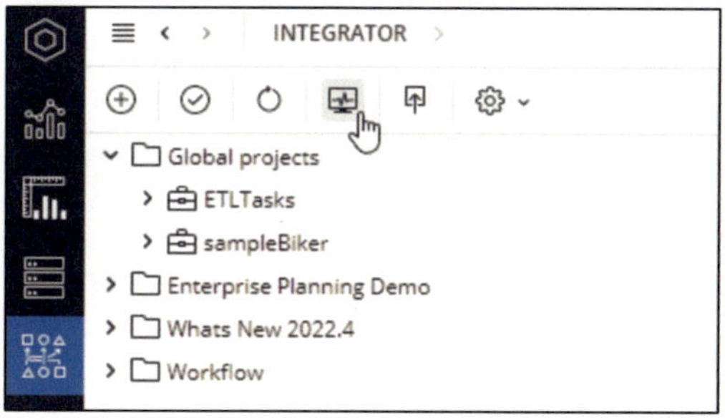

This will open the Integrator Monitor. Here you can review all the

Integrator logs. Look for anything colored orange or red. Orange is a warning (but the execution completes) and red means error:

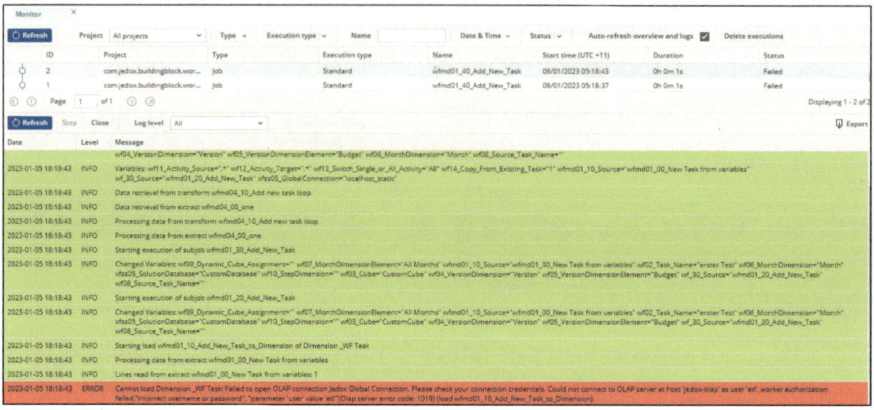

If you see a load or job that is failing, then this could be the cause of the error. There is a huge range of reasons for a particular load or job failure, but some common ones can include:

- **Authentication to source system failed** (e.g., username/password no longer valid). Has someone changed the password or removed the account?
- **Objects in the source system have changed or no longer exist**. The data that Integrator is expecting is not there. Has there been an update to the source system or a "clean up" which changed things?
- **A destination object in Jedox cannot be found** (e.g., Cube, Dimension, Attribute, Data Cell) has changed or no longer exists. Has someone recently changed something in the production system?

CALCULATIONS ARE NOT WORKING

Preflight Check

Since 2021, Jedox has a “Preflight Check” component for reports. This allows a report designer to define a set of value checks on a report and compare those saved values with the current numbers for the same intersection. This is valuable when migrating changed rules, integration processes or report calculations that affect report values from a development environment to production, for example. When executing the preflight check to check values, it will only tell you what cells have changed, not the actual reason for the change.

Are the rules activated?

Sometimes, if a target object (e.g., elements) changes, rules can deactivate themselves if the calculation is no longer valid. Try reactivating the rules by manually checking the checkbox:

Rule	Updated	Active
['Budget','10 Best Bike Seller AG','Turnover-Merchandise'] = B:PALO.DATA("Biker","Orders2",!'Y...	07/12/2010 03:55:39	☐
['Budget','10 Best Bike Seller AG','Cost of materials'] = B:PALO.DATA("Biker","Orders2",!'Years',!...	07/12/2010 03:55:46	☑
['Budget','20 Beta E-Commerce Ltd','Turnover-Merchandise'] = B:PALO.DATA("Biker","Orders2"...	07/12/2010 03:55:55	☑
['Budget','20 Beta E-Commerce Ltd','Cost of materials'] = B:PALO.DATA("Biker","Orders2",!'Year...	07/12/2010 03:56:01	☑
['Budget','30 Bike Trade Sarl','Turnover-Merchandise'] = B:PALO.DATA("Biker","Orders2",!'Year...	07/12/2010 03:56:11	☑
['Budget','30 Bike Trade Sarl','Cost of materials'] = B:PALO.DATA("Biker","Orders2",!'Years',!'Mo...	07/12/2010 03:56:19	☑

NOTE: Ensure the rule you are activating is one that should, in fact, be activated.

Supervision Server calculations

Supervision Server (SVS) calculations can be complex. By checking the server logs in the cloud console and filtering by Supervision Server, you should be able to see if there are any recent error messages in the logs that correspond to the time at which the system stopped working. You can then review any changes that may have been made to the SVS scripts from that point.

Have Attributes, Elements or Dimensions changed?

Rules can be driven off data, which could be stored in attributes for calculations. If the data changes, the rule value can change. In a data cell on a spreadsheet showing incorrect data, right click on a cell and select “Trace cell value”. This will show the rule used to calculate the cell (if there is one), and the specific inputs into that calculation for the selected intersection:

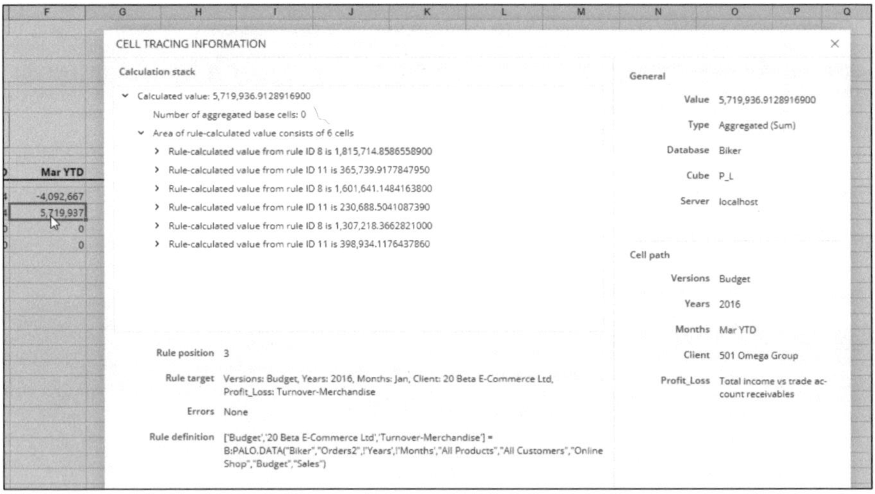

REPORTS NOT GENERATING

If reports are using the scheduler, check to see if it is activated:

If it is not activated and it should be activated, hover on the card and click Activate task:

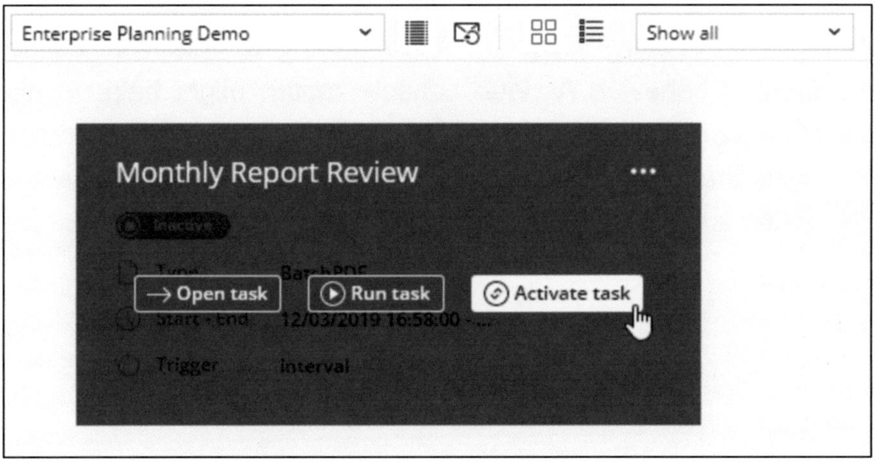

USERS CANNOT LOG IN

Has the Authentication changed for users? For example, a new Active Directory, changes in authentication protocols or new servers?

Large, long-lasting calculations

Sometimes, a large calculation routine (either rules or Integrator, or Supervision Server) can lock the system and prevent users logging in. Log in to the Cloud Console and check to see if resources (memory and CPU) are being used at or close to 100%. If this is the case, log into Jedox, go to Administration, Sessions. You can stop either a calculation thread or a complete session. If you are locked out and cannot log in, restart the Jedox instance from the cloud console.

Changes to Jedox cloud

If the infrastructure layer in Jedox changes, this can cause unexpected behavior. A cloud console restart might help. If this does not fix the issue, contact Jedox support directly. Remember to check the cloud console and the cloud status page before contacting support.

Final Notes

This book is intended as an introduction into the world of Jedox, an entrée to the platform and its capabilities. The Jedox platform is a powerful place to develop enterprise planning, reporting and analysis applications. What has been covered is just a starting point, with tips and advice to make your models and applications better designed and more efficient.

As your Jedox knowledge increases, you will start to see other areas within your organization where Jedox might be a great solution for often intractable problems. This is a common pathway for Jedox – it is great at solving business problems, especially ones that start in spreadsheets.

As with all technology, training and ongoing learning is the key. Learn as much as you can about the platform, through Jedox Academy courses, new version release notes or with the engagement of professional services from an authorized Jedox partner. Not all features are covered in this book – remember the Jedox knowledgebase is your friend! It often contains not only technical descriptions for specific components, but practical examples as well.

Best of luck on your Jedox journey!

About the Author

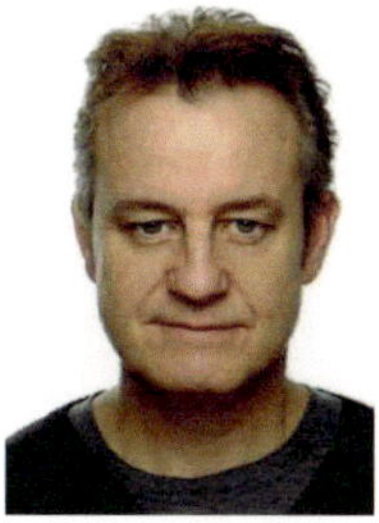

Chris holds a Bachelor of Business (Griffith University) and a Master of Fine Arts (Queensland University of Technology). He has spent 20 years in corporate performance management, using primarily in-memory technology. He co-founded Naked Data, an Australian based professional services company that was later acquired by Jedox. After a nearly 3-year stint in Germany as Director of Product management for Jedox, he now resides back in Australia.

Made in the USA
Las Vegas, NV
03 November 2026

57423447R00112